AF540418

LADAKH

(ECOLOGY AND ENVIRONMENT)

LADAKH

(Ecology and Environment)

BY

S.S. SAGWAL

Senior Scientist (Forestry)
S.K. University of Agricultural Science & Technology Srinagar
(Jammu and Kashmir) India

A.P.H. PUBLISHING CORPORATION
4435-36/7, ANSARI ROAD, DARYA GANJ
NEW DELHI-110 002

Published by
S.B. Nangia
A P H Publishing Corporation
7, Ansari Road, Daryaganj
New Delhi 110002
Ph.: 23274050
E-mail : aphbooks@gmail.com

ISBN 81-7024-433-1

2026

Rs. 1995/-

Printed at
Balaji Offset
Navin Shahdara, Delhi 110032

In the Everlasting Memory of My Grandfather

(*Chaudhary Bikhu Ram Sagwal*)

Preface

The book "Ecology and Environment of Cold Desert (Ladakh)" attempts to present, within a condensed form and in a readable fashion, a comprehensive detail of ecology and environment. Ladakh is the loftiest and remotest region of the Indian Republic. It is supposed that mother nature has been unkind to this region but it has many charming attractions. There are many agencies which disturb the ecology and lead to various disasters. Ladakh, long isolated from the hustle and bustle of the outside world due to its difficult topography and climatic condition, is now becoming an important centre of attraction of visitors - the historians, artists, scholars, tourists, educationists, nature's lovers, and many others who want to ward-off the beaten path. The ecological movement is becoming very popular all over the world and India is not untouched from its impact. However, in cold desert, public awareness and exposure to such outwardly trends is virtually non-existent. Therefore, Ladakhi people are confined only to their own world of wonders.

The Governments at Centre and State are very much conscious about the preservation and conservation of the heritage of this region. Now, several agencies are engaged in undertaking the work of development. There is an urgent need to improve the condition of down-trodden and the inclusion of environmental and ecological considerations are required to be placed immediately in the development planning. The bottom-up approach of planning should be encouraged so that the fruits/benefits of the developmental activities could be reached to the resource poor. For conserving the ecology and protection of environment of cold desert, priorities should be fixed to the problems associated with soil erosion, population pressure, degradation of land/forests, changing trends of fauna and flora, endangered species (i.e. plants, trees, wildlife), tourism, depletion of water resources etc.

For achieving overall development and prosperity of region it is essential to create awareness among the people to take steps for preserving the spoiling ecosystem. In case of negligence and carelessness, existence of man would even be in a great danger. The present book has been written with an object to get the people acquainted with several aspects of ecological and environmental principles. All the facts have been thoroughly taken care of and discussed in the leaves of different chapters of this book. It is earnestly hoped that the present book will prove a boon to the students, teachers, research scholars, scientists, planning-makers, foresters, administrators and several developmental agencies which are actively engaged in the region.

Dated: July 15, 1991

S.S. SAGWAL

Place: Kurukshetra.

Acknowledgements

The author is grateful to Prof. A. Ahmad, Vice Chancellor, S.K. University of Agricultural Sciences and Technology, Srinagar, Jammu and Kashmir State, India for encouraging to take up this work. I am highly indebted to my several friends, colleagues and the critics for their timely help. I am thankful to Dr. R.V. Singh, IFS, Director General (Retd.), Indian Council of Forestry Research and Education, Dehra Dun, (U.P.), India, Dr. H.G. Singh, Vice Chancellor, Rajasthan Agriculture University, Bikaner (Rajasthan), India, and Dr. P.K. Khosla, Director Extension Education, Dr. Y.S. Parmar University of Horticulture and Forestry, Nauni (Solan), H.P., India, who have always been source of inspiration for me. My sincere thanks are due to Dr. S.K. Chadha, Principal, Govt. Degree College, Kathua (Jammu), J & K State. I am highly thankful to my elder brother Chaudhary D.S. Sagwal, Manager, State Bank of Patiala, for his encouragement during the course of writing this manuscript. I have high appreciation for my wife Mrs. Savitri Sagwal (Guddi) for her unending enthusiasm, patience and immense help.

Further, I fully acknowledge M/s Madan Type College, Kurukshetra (Haryana) for typing and correcting this manuscript in a very short time.

S.S. SAGWAL

Contents

Chapter 1

Introduction

"The----------------------naturalist is one of the luckiest (persons) in the World. Winter or summer, rain or shine, walking or riding, his pleasures are near at hand. The great book of nature is open before him, and he has only to turn its leaves"

—*John Burroughs*

The Cold desert (Ladakh), has been regarded as the largest, the loftiest, and the remotest area of the World. But it has been placed in the sparsely populated regions in the Indian Republic (Chopra 1980). This has been the centre of attraction for the people of different walks of life not only from the Indian territory but the World over also. In this barren wilderness, nothing grows, not even the smallest shrub, hardly a blade of grass (Sahni 1990). The socio-economic structure of the area is paralytic and barren with limited natural, artificial and the productive resources. On the basis of physiography and other characteristics, this region has been designated as the roof of the World. It is really an enchanting view to see the peak (where the road passes) where it is written as, "Now you can talk to God". The ecological pattern is affected to a great extent due to some factors like scanty rains, extreme cold climate, droughts, xeric type of plants, arid environment etc. There are several salient features of this region on the analogy of which it is designated as versatile region inspite of inaccessible terrains and tracts. All these facts, accordingly, have been dealt in detail in the present text.

1.1 SITUATION

The cold desert is bounded on the north and east by China and in the north-west of Gilgit and Skardu. The districts of Baramula, Srinagar and Anantnag of Kashmir valley are on the west side whereas Himachal Pradesh is contiguous with its southern border. The region is situated in the inner-mountains of the Himalayas. The highest peak of this vast land touches 8,000 m (elevational height) whereas whole of this area is above

2,000 m. This physiographic region alone occupies nearly 70 per cent of the total geographical area of the State of Jammu and Kashmir. Geographically, the situational position as compared to J & K State is as follows (Table 1.1).

Table 1.1: Geographical position of cold desert

State/region	*Longitude (E)*	*Latitude (N)*
Ladakh	75°-15' to 80°-15'	32°-15' to 36°-00'
J & K	72°-30' to 80°-30'	32°-10' to 37°-00'

Source: SKUAST Background paper, 1983. After Bakshish, 1979.

1.2. DISTRIBUTION OF AREA

Cold desert, a constituent part of Jammu and Kashmir State, is situated in the extreme north-west part of the State with a geographical area of 95,876.00 km². It has marine origin because of its having been under a sea in the past (Burrand and Heydon 1907) which piled up huge sedimentary deposits until late tertiary period. It has very distinct features. It has now two districts viz. Leh and Kargil. The area of this region is as follows:

Table 1.2: Area (km²) in comparison with State*

District/State	*Rural*	*Urban*	*Total*
Leh (Ladakh) Kargil (Ladakh)	95864.70	11.30	95876.00**
J & K State	2,22236.00		2,22236.00

**Source*: SKUAST Background Paper, 1983.

** This includes 5180 km² area handed over illegally by Pakistan to China and 27,555 km² under illegal occupation of China. Also, this figure is of before 1947.

The geographical area of the State of J&K is equivalent to 6.74 per cent of the whole area of the country.

1.2.1. Physical Characteristics

Physically cold desert (Leh and Kargil of Ladakh area) and north-west plateaus of Gilgit (now under illegal occupation of Pakistan) fall in the inner-mountains. The prominent physical characteristics are as under:

1.2.1.1. ***Geology:*** The rugged terrain consisting of countless bare mountain peaks, converge into sandy valleys and are drained by the river Indus. The main geological series is the precambrain and consists of slates, phyllites, mica schist, quartzite, carbonaceous and graphite schist. Permo-carboniferous formation contains boulder beds, clays, lime stones etc.

1.2.1.2. ***Soils:*** The soils very from sandy to sandy loam and pure clay beds are also found at some places. The soils are neutral to slightly alkaline. The soil texture classes of some important places are *Leh* - Loamy sand (pH 7.8), *Zanskar* - sandy clay (pH 7.3), *Kargil* - loamy clay (pH 7.0) etc. (See Chapter 3). The major category of soils are cryobosols and lithic entisols (Murthy and Pandey 1978). The salt encrustations at slope from Kargil to Leh is, however, common, which is inimical to vegetative growth in early stages (Durrani *et al.* 1975). The soil pH ranges from 7.0 to 11.0 (Bhat 1965). The region on the whole is subjected to heavy pressure of western and north-westerly wind and its velocity goes sometimes as high as 100 km hr^{-1}.

1.2.1.3. ***River Systems:*** Indus; the principal river of Ladakh, locally known as 'Singgechhu' (Lion river). This river has two principal tributaries, viz. (a) Shyok, and (b) Nubra. In the region, Pangong, the largest lake, is 70 km long, 7 km wide and about 50 m deep. There are following valleys in the cold desert:

(a) Indus,

(b) Drass,

(c) Nubra,

(d) Suru,

(e) Zanskar.

1.2.1.4. ***Peaks and Mountains*** : Ladakh is famous for many peaks and mountains. Through these peaks and mountains, several passes are there where there is entry to Ladakh. The most famous is Zoji La Pass which is situated about 96 km north-east of Srinagar. This is situated at a height of 3,860 m in the Zanskar range of Himalayas. Following are the famous peaks and passes of Ladakh:

Table 1.3: Famous Peaks and Passes

Peaks/Passes		*Height (m)*	*District*
1.	Godwin*	8,611	Gilgit-Ladakh
2.	Gasherbrun*	8,066	Ladakh
3.	Mashashbrun*	7,821	"
4.	Harasnash*	7,397	Gilgit-Ladakh
5.	Ridden Peak*	8,068	Ladakh
6.	Broad Peak*	8,047	"
7.	Saserla	5,328	"
8.	Karakolam Pass	5,575	"
9.	Khardung La	5,602	"
10.	Tsaka La	4,724	"
11.	Lanak La	5,486	"
12.	Chorbata*	5,000	"
13.	Pensihal	4,410	"
14.	Zoji La	3,529	Srinagar-Ladakh
15.	Buril	4,199	Ladakh-Gilgit

Source : Gupta and Prabhakar, Jammu Tawi.

* Under the illegal occupation of Pakistan.

1.3. POPULATION

The population distribution of cold desert is affected by its physiographic, orographic, and stratigraphic set up. Following are the data of population:

Table 1. 4: Population as per Census (1971)

State/Region	*Persons*	*Males*	*Females*
Jammu and Kashmir	46,16,632	24,58,315	21,58,317
Ladakh	1,05,291	53,315	51,976

Source: Census of India, 1971. Part II (A) (i). p. 45.

Table 1.5: Population as per Census (1981)

State/District	Population	Males	Females
J & K State	59,87,389	31,64,660	28,22,729
Leh District	68,380	36,248	32,132
Kargil District	65,992	35,609	30,383

Source: Census of India, 1981.

1.3.1. Density of Population

Leh and Kargil are the only two important areas of settlements and density of population in Kargil is 5 persons per km² and 2 in Leh. Comparative density of population has been depicted (Table 1.6).

Table 1.6: Density of population (1981)

Country/State/District	Population/ sq. km.	No. of households per sq. km.
All India	216	31.79
J&K State	59	9.27
Leh District	2*	0.33*
Kargil District	5	0.80

Source : Census of India, 1981
* Density has been worked out on comparable area.

The rural population of Ladakh ranks highest among all the three regions. The data on rural population has been depicted (Table 1.7).

Table 1.7: Rural population of Ladakh (1981)

District	Rural	Urban	Total	Per cent Rural
Leh	59,662	8,718	68,380	91.36
Kargil	62,465	3,527	65,992	91.36

Source: Census of India, 1981.

Table 1.8: Rural population (per cent)

Region/Province/Valley	*Per cent of Rural Population*
J&K State	79.00
Kashmir Valley	89.87
Jammu Province	84.64
Ladakh Region	91.36

1.3.2. Sex Ratio

In Ladakh, the sex ratio is 879 per thousand of males (Census 1981). The literacy rate is very low. The Sex ratio is depicted (Table 1.9).

Table 1.9: Year-wise trend in sex ratio

					Female per 1000 males				
Region	*1901*	*1911*	*1921*	*1931*	*1941*	*1951*	*1961*	*1971*	*1981*
Ladakh	996	997	1029	1022	1011	990	970	975	879

Source: Gupta and Prabhakar, Jammu Tawi.

Another parameter of population is the religion. The ratio of different religions is quite contrasting. Table 1.10 shows the data of population based on religion.

Table 1.10: Population per cent according to religion

Religion	*Percent Population*
Muslims	44.60
Hindus	0.01
Buddhists	52.00
Rest	3.50

Source: Chadha, 1990.

1.4. ADMINISTRATIVE DIVISION

Ladakh has two districts, viz. Leh and Kargil. Kargil, situated on the Leh-Srinagar road and surrounded by the great Himalayan ranges capped by the Nunkun peaks, remains cut-off from Ladakh by heavy snowfall for nearly seven months of the year.

Now, Ladakh has 10 blocks and 72 panchayats. Both the districts of Leh and Kargil have 239 villages. The area of different towns of cold desert varies to a great extent (Table 1.11).

Table 1.11: Area and number of villages in Ladakh

District/Tehsil	*Area (ha)*	*No. of villages*
Leh	39,400	109
Kargil	14,292	105
Zanskar	4,023	25
Total	57,715	239

Source: Chopra, 1980.

1.5. FORESTS

The forests are the great assets of any nation. They have their vital role in the economy of the region/state/country. The importance of forests is peeping from the following lines,

> *"Forests are our green gold, and,*
> *sentinels of environmental defence"*.

1.5.1. Forest Area

Although, forests constitute a basic state resource of vital economic importance, there is no forest worth the name in this region. About 51.10 per cent area is under forest vegetation in J & K State, excepting Ladakh region, whereas it is negligible in cold arid zone. The distribution of forest wealth (Salaria 1979) in the State is as under:

Table 1.12: Distribution of forests

Region	*Forest Area (km²)*	*Percentage*
Jammu	12,012	45.51
Kashmir	8,976	56.56
Ladakh	466	0.00

It is evident from Table 1.12 that there are practically no natural forests in the region as a whole, excepting along the river banks and on moist alluvium. The tree species are poplars and willows found here and there.

1.5.2. Vegetation

The open sward mostly consists of *Lolium, Bromus, Sibbaldia, Arenaria, Taraxacum, Plantago* etc. Kargil is semi-arid with higher precipitation than Leh. Vegetational cover does not exceed 50 per cent in unirrigated areas (Misri 1980). In the moist soils and along the river bank, the vegetation consists of *Artemesia, Rosa, Corydalis, Euphorbia, Calamagerostis* etc. The tree species are *Salix daphnoides, Salix alba, Juglans regia, Populus deltoides, Morus alba,* etc. The natural vegetation (plantation) is in addition to 1,000 ha. of tree plantation (thee work of J&K Forest Department between 1956-80). Some of the medicinal and aromatic plants are also found in the area. The prominent species are, *Physochlina* spp., *Ephedra gerardiana, Podophyllum emodi, Dioscorea deltoidea, Saussuria lappa* etc. Sagwal and Bali (1987) have given elsewhere a detailed account of medicinal and aromatic plants.

Ecology is defined as a discipline which deals with the study of the relationships between animals, plants, and people and their environment. The ecological movement is getting much popularity all over the World. This has some impact in our country also. Similarly, environment means water, air, land, all plants, man and other animals living therein, and the interrelationships which exist among them.

The importance of ecology and environment should not be underestimated. It opens new vistas for getting knowledge about the cold desert. Therefore, full knowledge in detail will thus help in the all-round development of rural areas. In the following chapters, all earnest efforts have been made to explore the possibility to improve the existing conditions of the people and the environment.

Chapter 2

Climatic Considerations

The climatic considerations are the crucial factors to know about a particular place or region. These may include the following:

(a) Temperature

(b) Frost

(c) Snow

(d) Rainfall

The climate of Ladakh (cold desert) is supposed to be very dry and healthy. The climate of this region is viewed as:

2.1. TEMPERATURE

The temperature of a place depends upon four main factors viz. Latitude, longitude, rain-fall, and proximity to sea. All these factors influence temperature of cold desert in different ways. Champion and Seth (1968) noticed that the entire country including the high hills is bracketing in tropics as it experiences a mean annual temperature of 24°C or more. But in Ladakh, the mean January temperature may be seen as low as -13°C (Khan 1987). However, it is found upto -24°C which is extreme.

The intense type of cold is found in Rupsu, Rong, and some other high plateaus. The weather is, however, found milder in Central Ladakh and Nubra Valley. The absence of humidity helps to keep the atmosphere clear. The mean annual temperature is 40°F at Leh. The temperature rises from March to July and then shoots up sharply to about 63°F in the months of July and August. The Leh and Kargil of Ladakh are situated at about 3,800 m and 3,000 m respectively above the mean sea level, and therefore, they experience the extremes of hot and cold. The burning heat in the day may be followed by the piercing cold at night (Chopra 1980). There is a difference of 30° between the morning and afternoon (noted at Leh).

Table 2.1: The annual mean temperature of different places

Place	*Temperature (°F)*
Nubra	39
Leh	37
Zanskar	39

2.1.1. Effect of High Temperature

The high temperature has some of the mild effects. This effects the ecosystem in different ways. The effects may be attributed to.

(a) High temperature is harmful and even injurious to the physiology of the plant/vegetation. Some of the plants tolerate the range of temperature between 45° to 50°C. Death of plants may be due to the coagulation of some kinds of proteins.

(b) Dryness is also visible in the plants due to moisture deficiency. This is caused by very high temperature which attributes the heat up of the soil surface.

(c) There is a delicate relationship between respiration and photosynthesis which is generally disturbed at high temperature i.e. between 35° to 40°C.

(d) Cracking of stems in some plants may take place due to excessive heat.

These injuries can be prevented by natural adaptations; high sugar contents; deficiency of nitrogen. These factors contribute heat hardiness.

2.1.2. Effect of Low Temperatures

In cold desert (Ladakh), the temperature generally falls below the freezing point. This causes the chilling effect. Low temperature leads to the breakdown of more proteins. More fall in temperature may lead to two climatic phenomena viz.

(a) Frost

(b) Snow

2.2. FROST

As stated earlier that temperature goes upto -24°C which is the extreme case. In Ladakh, temperature remains below freezing point for a good period of time. Frost occurs due to the cooling of air to the temperature below the freezing point. It is of the following types:

(i) *Radiation*: This occurs when the nights are clear. A rapid cooling of air layers takes place just at the surface of earth. A rapid loss of heat by radiation may cause this frost.

(ii) *Advective*: This is caused by the outside air and gets cooled in the region. This occurs generally in the form of pockets, locality or the holes.

(iii) *Pool*: These are the characteristics of valley areas. The cold air gets accumulated and the temperature goes down considerably causing pool frost.

2.2.1. Damages due to Frost

A large number of damages and injuries are caused by frost in the region. Following damages may be caused:

(i) young plants are affected and may get killed due to frost.

(ii) Frost lifting is caused sometimes. The seedlings are lifting above the ground/soil.

(iii) Dieback of crowns and the saplings.

(iv) Frost cracks and cankers may also be seen.

2.2.2. Protection Measures

The following measures should be taken:

(a) Watering early in the morning during winter season and at the evidence of frost.

(b) Erecting the smoke screen in the affected areas.

(c) Raising of nurse crop a bit early than the main crop.

(d) Weeding should be deferred for sometime as they protect the other crops to lessen frost effect.

(e) Adopting suitable system of silviculture, if, possible in the newly raised crops, e.g. shelterwood and selection system.

2.3. SNOW

The Snow occurs in the areas of higher elevations, where lowering of temperature leads to condensation of water vapour present in the atmosphere - in the form of snow. The amount of snowfall received in a particular region/area is mainly dependent upon:

(i) Topography,

(ii) Altitude/elevation,

(iii) Amount of winter precipitation, and

(iv) Climate of area (temperature etc.)

2.3.1. Cold Desert - A House of Snow

The altitude above which snowfall occurs, varies to a great extent from place to place. In Western and Central Himalayan tracts, the lowest limit of annual snowfall is around 1,400 m elevation and the line of perpetual snow or snowline is supposed to be around 2,000 m. Some places of Ladakh are also situated on height (a.m.s.1.) of 3,800 m even more. However, the intensity of snowfall varies to a great extent in different parts of the cold desert. It may vary from few centimetres to metres. Sometimes 18 to 20 feet of snowfall are also experienced.

2.3.2. Effect of Snow on Vegetation

The effects of snow/snowfall can be categorized into 2 groups viz. Quantitative and Qualitative.

2.3.2.1. ***Quantitative Effects***: The quantitative effects of snow are observed in various ways.

2.3.2.1.1. *Mechanical bending*: The snow causes a mechanical bending of plants particularly at the young stage. When snowfall occurs, it accumulates on branches/crown of plants/trees. It has been observed that even the sapling and pole stages of trees are worst affected by this process of accumulation. Therefore, some weight is experienced by the plants/ trees and results in bending of stem and even the branches. Generally, bending has been noticed towards the base and when the trees mature, this can be more visual. Due to this bending characteristic, the value of trees is reduced to a considerable extent.

2.3.2.1.2. *Breaking by settling forces*: The young stages of trees are worst affected. The young trees are broken by settling forces of snow under which they are buried. The creeping pressure of snow is in the proportion of the depth of snow. The seedlings in the nursery are mostly the victims of such forces. For example, Poplars, Willows, Ash trees etc. are worst affected.

2.3.2.1.3. *Breaking through Crowns*: The branches and the tops of trees and shrubs may be broken up by the weight of accumulated snow. The crowns retain the snow and ultimately the crowns are severely affected which result in the breakage. For example, the crowns of willows and poplars are most susceptible to the damage by the snow-break.

2.3.2.1.4. *Snow Slides/avalanches*: The snow sliding and avalanches affect the vegetation in a number of ways. Some of the ways are as follows:

(i) Excessive erosion of the ground floor. The weight of snow on the forest floor or the ground exerts pressure which results in the cracks and loosening of the soil/earth. This, therefore, leads to erosion and washing away of the ground.

(ii) Uprooting of trees: The sliding processes and avalanches uproot the trees/plants from the snow prone areas. The trees are liable to fall down irrespective of the stages of roots of trees (may be seedling/sapling/pole) or the tree itself. This results in the loss of forest vegetation and affects the ecosystem.

(iii) The small strips devoid of vegetation are created along the path of snow slides. The snow slides/avalanches come from a very long height which exert a heavy force. This force, of course, breaks the trees, shrubs or other vegetation. Thus, the small strips which are generally devoid of any vegetation are created. For example, outskirts of Sonamurg mountains.

(iv) In depressions, where snow slides occur frequently, the large areas are without a vegetative cover. This is most probably due to the reason that the earth and stones/pebbles they come down through sliding processes and accumulate in the depressions resulting in the burrying of vegetation.

Hence, quantitative effects are more severe and should be looked for. This results in the reduction of the vegetative cover.

2.3.2.2. *Qualitative Effects* : There accounts large number of qualitative characters of the vegetation which are extremely affected by the snow in the region of cold desert. Following are probable qualitative effects :

2.3.2.2.1. *Reduced Growing Period* : The snow-bound areas are regarded as temperate, cold desert, sub-zero area, depending upon the intensity of winter, snowfall etc. The day length becomes shorter in such regions and with the commencement of winter, the growing period gets reduced to a great extent. Most of the species start shedding their leaves too early in comparison to other type of areas (less snowfall). Further, the sprouting of leaves and flowers etc. starts very late in snow-bound areas and this reduces the growing period. For example, many species of Poplars, Ailanthus etc. hence, growing cycles become limited due to the occurrence of snowfall.

2.3.2.2.2. *Discoloration and fungi* : It has been seen that many plants/trees lose their original colour of bark. However, during this period, many a number of trees shed their leaves, whereas, others do not do so but certainly change their colour of leaves and branches. The favourable conditions are created due to the effect of snowfall which give birth to certain fungi. Consequently, a large number of plants are liable to the attacked and reduce the timber value. For example, *Fomes badius*, *Trametes* spp. etc. sometimes these fungi become fatal to the plants/trees.

2.3.2.2.3. *Respiration/photosynthesis* : The respiration becomes slow in the presence of snow. This may be due to low pace of intake of Carbon-dioxide (CO_2) and release of oxygen (O_2) and consequently results in the creation of environmental pollution. Also, during the process of photosynthesis Carbohydrates (CHO) are manufactured from CO_2 (Carbon dioxide) and water (H_2O) in chlorophyll containing tissues of plants exposed to light. Several conifers become somewhat chlorotic during and at the time of snowfall. This is most probably because the destruction exceeds synthesis of chlorophyll at very low temperature intensity at the time of snowfall (may be heavy). Also, sunlight intensity is not adequate during snowfall.

2.3.2.2.4. *Other Effects* : In addition to above effects, snowfall has a severe affect on other activities of vegetation. Some of the visual effects are as follows:

(a) Phenological;

(b) Pollination;

(c) Enumeration difficult;

(d) Debris removal etc.

2.3.3. Control Measures

Though the effect of snowfall cannot be controlled fully, yet some preventive measures can, however, certainly be employed to reduce the effect to a great extent. Some of the possible measures could be:

2.3.3.1. ***Polyculture***: The deciduous plants can make themselves best associates of conifers, e.g. *Populus ciliata* in the forests of *Cedrus deodara*. Therefore, polyculture in the cold desert must be given preference over monoculture to reduce the insect-pests and diseases hazards and consequently the plantation loss. The transition from conifers to deciduous vegetation/forests is very wide, so wide in fact that many authorities recognise it as a distinct formation type, often called as Ecotone Mixed Vegetation.

2.3.3.2. ***Regular Thinnings***: These are the great assets for the insurances of maintaining strong plants as the weak, dying, diseased and moribund trees are removed. Therefore, light thinning, if at all, be always preferred over heavy ones due to loss of lops and tops. Thinnings make room for non-accumulation of snow on the branches, leaves, and in crowns itself and the vegetation is saved to a great satisfaction. In this way, the trees become strong enough to bear the onslaught of snow damages.

2.3.3.3. ***Working under Selection System***: The forest vegetation should as for as possible be worked under the selection system. The individual plants are removed depending upon the forester's advice as to what is the best for stand development. The criteria used in selecting trees to be either released or cut include crown class, vigour, spacing, form and branch characteristics.

2.3.3.4. ***Choice of tree species***: It plays a vital role in saving the vegetation. Therefore, those species should be selected for the plantations which can bear the effect of snowfall to some extent. Moreover, fast growing species may be well suited as they can complete their rotation at the shortest possible period and can be exploited easily.

There may be several other ways to tackle the problems of the effect of snow on vegetation. However, there is a great need to do some more researches on this aspect.

2.4. RAINFALL/PRECIPITATION

The cold desert is the innermost line of high mountains and the physiography of this region is extremely rugged, formidably arid, frost bitten and mostly devoid of any significant vegetation. The area remains at below zero for about five months (November to March). The annual precipitation is about 160 mm (Kangoo 1989). This is presented in Table 2.2.

Table 2.2: Precipitation (rainfall) data of cold desert (Ladakh)

Month	*Precipitation (mm)*
January	10.60
February	76.00
March	8.90
April	6.10
May	5.98
June	4.80
July	12.70
August	16.50
September	9.10
October	3.10
November	1.00
December	5.00

However, the summer rainfall of cold desert is 43.10 mm. It is evident that rainfall is scanty in the region. This leads to the setback to the vegetation growing there. Rainfall affects the ecosystem in a number of ways. These are viewed as under:

2.4.1. Damages caused by High Rainfall

Rainfall in excessive amount and scanty nature causes a number of damages. Following are the probable damages caused by high rainfall:

2.4.1.1. ***Reduction of vegetative cover***: It is seen that a large amount of area is lost due to heavy rainfall. But in cold desert this is not, however, a big problem. The only thing is that runoff water may cause damage of vegetative cover.

2.4.1.2. ***Debris deposition*** : Due to more precipitation (occurs comparatively in some month i.e. February), the debris is carried out/transported and burried in some places (Negi 1983). However, this is not a salient/ regular feature.

2.4.2. Damages caused by Low Rainfall

It is the main attraction that cold desert remains dry for most of the time. It has been estimated that about seven months, the area of cold desert remains dry which poses a great threat to the ecological balance. The damages caused by scanty rainfall may be as follows:

2.4.2.1. ***Drying up of the river*** : It is apparent that due to lack of rainfall, the area of cold desert goes dry which ultimately leads to the drying up of the river. This phenomenon creates a number of problems viz. fishing is harmed, lack of drinking water to the animals (both of domestic and wild ones) etc.

2.4.2.2. ***Cracks in the soil*** : The excessive dry condition may lead to the more heat which causes, sometimes, the cracks in the soil which are harmful to the crops and the micro-organisms.

2.4.2.3. ***Wilting of plants***: The low rainfall does not meet the requirements of water for the growing plants. If, proper rainfall is not received by the plants they may get wilted which is a server setback to the growing/ coming up vegetation in the area.

2.4.2.4. ***Mars the sowings*** : For a successful sowing operation, moisture is required in the soil. In the absence of rains/very less rains, this operation is badly affected which is fatal to agriculture, pastures, forestry etc.

In conclusion it can be said that climatic considerations are of vital importance for cold desert. The climatic factors affect the ecosystem in a number of ways. High and low, both types of temperatures are met with. They affect the area in different ways which influence the environment. Frost biting also takes place in Ladakh which is a threat sometimes to the growing vegetation. Similarly, snow has its effect on the vegetation. Rainfall is a crucial factor in the region as it affects the vegetation as well as the wildlife. Therefore, climatic factors affect the ecology of cold desert in several ways.

Chapter 3

Soil and Landuse

"Food comes from the earth. The land with its waters gives us nourishment. The earth rewards richly the knowing and diligent but punishes inexorably the ignorant and slothful. This partnership of land and farmer is the rock foundation of our complex social structure".

—W.C. Lowdermiek

Soil is a medium for plant growth. It is a complex natural material derived from disintegrated and decomposed rocks and organic materials. This provides nutrients, moisture and anchorage for land plants. The following statement holds good for the importance of soil.

"Soil sustains plant;
Plant sustains life".

Soil has four main components which are supposed to be the essence of life of not only of the plants but the human-beings also. These components are as follows:

(a) Mineral materials,

(b) Organic matter,

(c) Air, and

(d) Water.

All the four components are found widely in the combined form but in varying amounts in different type of soils.

3.1. FORMATION OF SOIL

The development of soil is a long process. This involves both physical and chemical weathering. However, this is integrated with biological activities. The soil formation factors include a long series which are influenced in a number of ways. Following are the most accepted factors for this process (CFA 1985):

3.1.1 Parent Material

This is the most important component. It refers to the material from which the soils were formed.

3.1.2 Climate

It consists of two prominent components which influence the place/site. The major components are,

(a) temperature, and

(b) moisture.

3.1.3. Living Organisms

These include microscopic and macroscopic plants and animals.

3.1.4. Topography

This represents the following,

(a) shape, and

(b) position of land surface.

3.1.5. Time

It is the period during which parent materials have been subjected to the soil formation.

3.2. PHYSIOGRAPHIC UNITS

The region of cold desert (Ladakh) has a typical topography which consists of conical hilly peaks with clusters of spurs. This leads to the formation of V-shaped valleys. Ladakh is situated at the north-western end of the Himalayan range. It makes the lofty lands of the area. There are following *three* physiographic units of the region :

(i) High hills,

(ii) Pediplane-foot and toe slopes of alluvial and glacial moraines, and

(iii) Valley - V-shaped narrow inhibited by the people with some cultivation/vegetation near the water courses.

The soils are generally sandy. These are derived from debris of weathering rocks. However, these are influenced by great diurnal as well as the seasonal alteration of temperature. This leads to the disintegration of the rocks.

3.2.1. Agencies of Weathering

Soil formation is the resultant of weathering of rocks. There are several agencies which influence the weathering in cold desert. A few, of course important, are as follows:

(i) *Temperature*: The temperature variation causes physical weathering in Ladakh. However, this is of intense nature.

(ii) *Rainfall*: Rainfall is a crucial factor of climate. Since, rainfall is low in the area, so chemical weathering is sure but slow in pace. However, the mineral resources are of high rank.

(iii) *Penetration of water*: Generally the rain water penetrates in the soil which helps in accelerating the weathering process. However, the brittle crust reduces the effective rainfall.

3.3. SOIL TEXTURE

Texturally, the soils of cold desert are recognised as 'Coarse'. The range of sand in the soils varies from 20 per cent to 68 per cent. The soils are thus of loose type which have a good aeration provision. This leads to lack of capacity to absorb and hold sufficient moisture and nutrients. The mechanical analysis of soils of different places of Ladakh region has been depicted in Table 3.1.

Table 3.1: Mechanical analysis of soils

Place	*Course sand (%)*	*Find sand (%)*	*Silt (%)*	*Clay (%)*	*Texturel class*
Drass	0.88	19.82	53.90	25.00	SC
Kargil	6.44	61.10	3.30	20.50	LC
Leh	4.17	24.83	2.50	8.50	LS
Suru	1.00	58.55	28.50	12.00	L
Zanskar	3.77	54.82	15.00	25.50	S.C

Source: Department of Agriculture, J & K State.

SC = Silty Clay; LC = Loamy Clay
LS = Loamy Sand; L = Loam
S.C = Sandy Clay;

It is evident from Table 3.1 that silt varies from 2.50 to 53.90 per cent and the clay from 8.50 to 25.5 per cent. Talib (1986) observed sand content in the soils of Kargil (Khurbatang plateau) as relatively less (51.29 to 93.00 per cent) with sandy loam as predominant surface texture of the soils.

3.4. CHEMICAL CONSTITUENTS

The chemical constituents show very interesting facts and information. The organic matter varies from 0.01 per cent to 1,56 per cent. Most of the soils show less organic matter which is due to lack of vegetation in the area. This exhibits poor microbial activities and the resultant is low humus. The highest constituent of acid insoluble residue varies from 73.85 to 87.33 per cent. Table 3.2 shows the chemical constituents of soils of different places of Ladakh.

Table 3.2: Chemical Constituents of the soils of different places

Constituents (%)	*Leh*	*Kargil*	*Zanskar*	*Saspol*	*Chang-thang*
Loss of ignition	3.56	2.73	1.80	5.74	4.55
Acid insoluble residues	73.85	87.33	84.63	85.50	86.88
Sesquioxides	12.00	12.18	7.50	9.80	6.80
Ferric oxide	6.20	7.00	4.30	2.90	1.50
Calcium oxide	2.90	2.00	3.70	0.30	1.20
Magnesium	0.50	1.80	0.04	0.02	1.10
Phosphorus	0.18	0.12	0.25	0.21	0.19
Potash	0.01	0.03	0.03	0.01	0.01
Nitrogen	0.14	0.18	0.07	0.05	0.03
Carbon	0.01	0.01	0.67	0.69	0.33
Organic matter	0.01	0.02	1.56	0.68	0.56
pH	7.80	7.00	7.30	7.50	8.50

Source: Department of Agriculture, J & K State.

Organic carbon content of the soils of Leh is low (0.11 to 0.35per cent) which is attributed to lack of vegetation cover, poor microbial activities and low humus (Talib 1986). He further observed that high temperatures are favourable to rapid mineralization of organic residues and hence the organic matter is low. From Table 3.2, the data show that pH values range from 7.00 to 8.50 which indicates the alkalinity status of the soil. There is one interesting fact about the soils of Ladakh that the boulders and pebbles are found invariably at different depths of profiles.

3.4.1. Soil Classification

The soils of cold desert are classified as 'High Altitude Desert Soils'. However, these soils may be grouped under 'Cryorthents'. This is due to the fact that the region comes under the Cryic temperature regime.

Based on the colour chart of the soils, the cold desert soils exhibit mostly the light grey (10YR 6/1) to greyish brown (10YR 5/2). However, these soils are regarded as structureless. Also, the soils can be grouped differently depending upon the colour of soils of different places. The above figures show rough estimate which can be assumed as the base for further knowledge about soils.

3.5. LAND CAPABILITY CLASSIFICATION

The knowledge of land capability classification is very essential before going for any developmental activities in the region. The land is classified into various categories on the ability of the land to produce on a sustained basis (Anon. 1985). The term land oftenly includes soils, minerals, climate, water supply, location in relation to markets, vegetative cover and improvements. Broadly speaking, the land is classified into following classes:

3.5.1. Class I

The land of this class is most suited to intensive cultivation. It poses very less limitations and the land can be put to all types of uses. It has easily workable soils and is not liable to abnormal erosion. However, under some specific circumstances, it requires fertilizers, organic manure, lime etc. As a whole, this class is most versatile which is put under many uses.

3.5.2. Class II

The soil of this type of land poses a few limitations. It has moderate type of slopes. Soil is deep and liable to very little erosion. However, some times the limitations may require some remedial measures. A few and prominent measures could be, (a) terracing, (b) strip cropping, (c) crop rotations, (d) controllable irrigation facilities, (e) stubble mulching, (f) organic manures etc.

3.5.3. Class III

This class poses somewhat greater limitations for its use. The land can be used for cultivation by adding some amendments. This category shows steep slopes. The salient features of this category are,

(a) moderate to steep slope,

(b) moderate to high susceptibility to erosion,

(c) sandy to very sandy soils,

(d) shallow soil depth,

(e) low fertility gradient etc.

3.5.4. Class IV

Poses very severe limitations and can be best utilized for a perennial vegetation cover. However, it can be used for cultivation occasionally. The cropping is restricted due to

(a) adverse natural features,

(b) unfavourable climatic conditions. Measures against erosion problem must be taken up.

3.5.5. Class V

The land is not suitable for any type of cultivation. This land is most suited to forestry purposes and grazing. The cultivation is impossible due to several reasons which are as follows:

(a) severity of wetness and dryness,

(b) stoniness condition,

(c) moderate to steep slope,

(d) moderate to shallow soil depth etc.

3.5.6. Class VI

Not suitable for cultivation but good (has a few limitations) for forestry and grazing/pastures purpose. Some measures for corrections can be taken up as,

(a) deferred grazing,

(b) rotational grazing,

(c) contour furrow/ checkdams etc.

The land exhibits the following characteristics, viz.

(a) too steep,

(b) too stony,

(c) too dry or too wet etc.

3.5.7. Class VII

This is also not suited to cultivation and are usually rough, very steep, very dry or wet. This has severe limitations for its use to grazing/pastures or forestry. However, this class can be made suitable for grazing by amending. Following are the measures to be taken,

(a) closing the area to grazing.

(b) prohibit lopping.

(c) adopt contour furrows, ridges, etc.

(d) planting suitable tree species etc.

3.5.8. Class VIII

The land is unsuited to cultivation. This is also not used for forestry or grazing/pastures purposes. However, this land can be put under the use of,

(a) wildlife conservation,

(b) watershed management,

(c) recreation purposes etc.

The need for land use planning is frequently brought about by changing the requirements and pressures involving competing uses for the same piece of land. This helps to guide decisions in such a manner that environment resources can be put to maximum use for man (Vijayalakshmi 1986). The best way to protect would be to put the land under pasture or forest. However, the local demands of the people is to be kept in mind but there is no alternative. For getting maximum benefits, all possible land must be brought under social forestry. This will pave the way to harness every possible use.

3.6. SOIL AND WATER CONSERVATION

The physical weathering is more visible and prominent than that of chemical due to dominance of physical weathering agents. Soils are found as shallow in depth as stones are generally met after 35 cm in most of the soil profile (Handoo *et al.*1984).

3.6.1. Management Strategies

It is very essential to evolve some of the management strategies to tackle the problem. Hence, an integrated approach of crop husbandry, agroforestry and horticulture. The following strategies must be adopted:

3.6.1.1. ***Soil reclamation***: This can be done in two ways, viz.

(a) Levelling: The project area must be cleaned of gravels, pebbles, cobbles, stones, etc. Thereafter, land must be levelled to the possible extent. This is used for uniform irrigation purposes.

(b) Contour bunding: This must be operated for the areas having 2-3 per cent slope. However, the terracing is performed for the areas exhibiting more than 3 per cent slope.

3.6.1.2. ***Water harvesting***: There is regular flow of streams during season. This water comes from the glaciers and it is the unchecked flow. Water comes with great pressure from mountains, so it causes gully erosion due to the force. This water can be harvested judiciously. For this purpose check-dams can easily be constructed and flow of water can be directed towards the tanks (storage tanks). From this crucial point (storage tank) this water can be used for irrigating the fields. This can be done throughout the growing period i.e. from May to September. This water can be harvested economically as and when needed.

3.6.1.3. ***Use of organic matter***: Organic matter can be used for increasing the efficiency of soils. In Ladakh most of the uncultivated lands are without natural vegetative cover excepting a few pockets of pastures. The functions of organic matter are manifold. This performs,

(a) smooth infiltration of water,

(b) reduces surface run-off,

(c) binds soil particles,

(d) prevents soil erosions,

(e) increases water holding capacity of soil,

(f) source of energy for soil micro-organisms,

(g) reduces alkalinity etc.

Therefore, system of crop rotation must be adopted. The organic matter could be added through green manuring. The stubbles are transformed into humus with a span of time.

3.6.1.4. ***Anti-erosion technique***: Wind-erosion is the main problem. This can be reduced by adopting integrated approach of horticulture and agroforestry. The use of fruit trees and xerophytic shrubs/bushes can be intercropped along the bunds/boundaries. This serves the following purposes, viz.

(a) This will serve as wind breaks and reduces wind erosion.

(b) This improves the environmental system and raises sources of income of the farmers.

3.6.2. Conservation Measures

The soil and water conservation measures adopted for the development of cold desert (Gupta 1986) are,

3.6.2.1. ***Farm ponds***: This is a new technique for conserving soil and water. In this technique, a pond (the number of ponds depend upon the size of the farm i.e. it may be more than one also) is constructed on the farm itself. This is used for storing water and helps soil from its erosion. The water, thus, stored can be utilized judiciously.

3.6.2.2. ***Gully erosion control***: It has been seen that most of the population of the region is confined to the vicinity of nullahs. This is apparently true as these are the sources of water both for humans and the crops. The best way for controlling gully erosion is the spurs and check-dams of registered dimensions.

3.6.2.3. ***Bench terracing***: This is also a viable method generally employed for soil and water conservation. The practice is adopted in the areas of slopes of even 20-50 per cent. These benches can easily be used for cultivation of crops and heavy use of commercial non-traditional inputs (fertilizers) and the manures (the farm yard manures and the green manures).

3.6.3. Research Programmes

For the over-all development of the region, research is very essential for evolving some good technologies which can easily be used. Therefore, research should be geared up immediately. The main research strategies could be as follows:

3.6.3.1. ***Reclaiming alkali soils*****:** Research is needed to evolve the right quantities of soil amendments for reclaiming such areas. The selection of suitable tree species for such soils should be carried out by conducting appropriate researches. Also, micronutrients e.g. zinc, iron, should be evolved for their suitability to crops in such soils.

3.6.3.2. ***Soil testing*****:** Soil testing is the need of the hour as this has been lacking in past. But now, with the introduction of fertilizers, it is felt very essential to test the soils before making any recommendations of fertilizers for a particular crop/soil. Therefore, studies must be taken up earnestly to work out relationships between soil tests for nutrients by different methods and crop responses. This will help in boosting the yield of crops for optimum levels and economic responses.

3.6.3.3. ***Micronutrients*****:** There is an urgent need to work-out the status and requirements of micronutrients of different areas of Ladakh. This should be done for different crops.

3.6.3.4. ***Soil Survey*****:** Soil survey should be carried out throughout the region. The soil survey map should be prepared. This will help in classifying the soil and land irrigability classes for efficient use of water. (Khan and Wani 1986).

3.6.3.5. ***Miscellaneous studies*****:** Some more studies should be taken up on the following:

(a) improving the nutritional status of the pastures.

(b) use of bio-fertilizers for Nitrogen-management of soils,

(c) pedology and minerology of soils,

(d) relationship of soil characteristics and the efficient use of water,

(e) windbreaks and their effect on the soils.

From the above points, it is clear that research is needed for several components. If, this effort is carried out in true spirit, then water and soil conservation can be adopted easily and with very less expenses with the local available know-how. This will pave the way for further improvement.

3.7. CONSTRAINT ANALYSIS

It is evident that soil and water conservation is the need of the day in cold desert as it affects the crop production to a great extent. A large number of constraints are in the way of sustained crop production in Ladakh. Following is the probable critical analysis of constraints:

3.7.1. High Wind Velocity

The area is prone to the high wind velocity. The high frequency of wind blow causes soil erosion and brings many problems. Thus, high wind velocity poses a great constraint.

3.7.2. Loosening of Soil

In this region, loose stones, cobbles, pebbles, gravels etc. are found in abundance. These are scattered over the surface of land. These are the hinderances in the way of soil and water conservation.

3.7.3. Poor physical conditions

The area experiences the higher rate of infiltration and percolation. Also, shallow skeleton soils are met within the region. Mostly the soil texture is of sandy nature which otherwise is of no use. This type of soils are liable to be affected by several agencies.

3.7.4. Irrigation Facilities

The whole of water is impossible to use for irrigating agricultural crops. This is generally attributed to,

(a) low amount of precipitation,

(b) deep level of water in the rivers which cannot be utilized for irrigation.

3.7.5. Miscellaneous

A large number of constraints are in the way of crop production related to soils. A few of them are as under (Wani *et al.* 1977):

(a) low organic matter,

(b) alkaline soils, and

(c) low status of fertility.

In conclusion, it can be stated that mostly the sandy soils are found in the whole tract of cold desert. The soil formation is affected by several factors, viz. parent material, climate, living organisms, topography, and the time. The whole of cold desert has been divided into three physiographic units. The organic matter indicates the low content which exhibits the less/reduced microbial activities. The PH values show that the soil trend is towards the alkalinity. The soils of the region have been grouped in 'High Altitude Desert Soils'. The land capability classification has been critically discussed. The soil and water conservation has also been discussed with all the possible measures. This includes the management strategies, and measures to tackle different problems. Some suggestions for research activities have also been underlined for overall development of cold desert. The soil and water faces many constraints for crop production. Therefore, soils should be managed by chalking out the viable media so that the enriched source ((soil) can be fully utilized for the prosperity of the region.

"A fertile soil is not always a productive soil;
But a productive soil is always a fertile soil".

Chapter 4

Agriculture

"There are two spiritual dangers in not owning a farm. One is the danger of supposing that breakfast comes from the grocery, and the other that heat comes from the furnace."

—*Aldo Leopold*

Ladakh has been the centre of attraction of many people from the time immemorial. It has attracted the people like biologists, geologists, hydrologists, tourists, scientists not only from India but the world over also. Many opportunities like altitude research, studies of ecology of fauna and flora, grasslands, forest resources etc. are offered by this region. Inspite of all these blessings of mother nature, the people are plunged into the jaws of poverty which is the worst evil of the society in this hilly tract of Himalayas. The economy of the region is in deplorable conditions. Therefore, there is an urgent need to pay attention towards the economy and social set up. The standard of living depends upon agriculture as it is the main stay of the people. An attempt has been made in this text to explore the possibility of creating a sound agriculture environment in the region so that it can be made firm-footing. This effort will help in meeting out the different requirements of the people.

4.1. LAND HOLDINGS

Land holding is a crucial factor in adopting agriculture. The condition of land holdings is poor due to small size and the limitations posed by land constraints. The data on land distribution indicate that more than 60 per cent of holdings fall between the range of less than 0.5 and 1.0 ha and more than 78 per cent within the maximum limit of 2.0 ha (Table 4.1). The average holding size is 1.57 ha and only 21 per cent of the total area is under tenancy system.

Table 4.1: Farm size distribution

Size Class (ha)	*Land holdings (per cent)*
<0.5	41.19
0.5-1.0	19.11
1.0-2.0	18.37
2.0-3.0	10.77
3.0-4.0	5.03
4.0-5.0	2.35
5.0-10.0	2.65
10.0-20.0	0.36
>20.0	6.17

Source: University Background Paper, 1983.

4.2. IRRIGATION FACILITIES

It is evident that the region has an extensive surface water system but cannot be tappled due to extreme type of problems. The net irrigated area, however, comes out to be 100 per cent through Government canal sources.

4.2.1. Irrigated Area by Sources

The area is 100 per cent irrigated through the sources (Table 4.2). The whole estimated area which is irrigated is supposed to be 18,000 ha.

Table 4.2: Irrigated area by sources

Source	*Irrigated area ('000 ha)*	*per cent of net irrigated area*
Government Canals	18.00	100.00
Tanks and Wells	N/A	N/A

4.2.2. Other Sources

The area is not smooth at all and it is not possible to irrigate the area for agricultural purposes. However, people (farmers) make use (on small scale) of glaciers' water by constructing some tanks and small dam like structures. Sometimes the retaining walls are erected to divert the flow of water. But there is no definite information as how much area is covered under this self-made effort of the farmers. This is the sound reason that people are settled down in the vicinity of some water courses so that water can be tapped with the help of local resources.

4.3. CROPPING PATTERN

As a whole, agriculture is the chief occupation of the people in J&K State covering about 65 per cent the total population.

According to the census of 1981, rural population of Ladakh was 91.45 per cent. It is estimated that more than 80 per cent of population (rural) is engaged in agriculture profession in one or the other form. Therefore, agriculture is the main occupation in cold desert. The socio-economic structure is weak due to limited cultivation opportunities. The cropping pattern is viewed as follows:

4.3.1. Agricultural Crops

The main cereal crops which are cultivated in the region are, wheat, grim, pulses, small millets etc. The area under different crops is depicted (Table 4.3).

Table 4.3: Area under cropping pattern

Crop	*Area (ha)*
Wheat	3,804.00
Cereals and millets	10,035.00
Pulses	860.20
Other Crops	4,052.00
Total area	18,752.00

The small millets, due to their wide adaptability, varying sowing seasons, and ability to grow under poor conditions which constitutes dominant food grain crop, occupies 53.51 per cent of the total cropped area. The productivity among food grains is the highest in case of high yielding wheat and local wheat, averaging 14.00 and 10.00 q ha^{-1} respectively followed by grain and pulses with an average of 8.00 q ha^{-1} in each case (Table 4.4).

Table 4.4: Yield of major crops (Av. basis)

Crop	*Yield (q)/ha*
High Yielding wheat	14.00
Local Wheat	10.00
Grim	8.00
Pulses	8.00
Small millets	7.50

4.3.2. Horticultural Crops

Cold desert is also famous for horticultural crops. These crops include apple, apricot, grapes etc. The cold desert is devoid of any good vegetation. However, a few patches of lush green meadows are seen scattered in whole region. Some patches of stunted cedars *(Cedrus deodara)* and Willows (*Salix* spp.) can also be seen here and there. The fruit production is 15.00 q ha^{-1} (average), which include apricot (fresh and dry) and the apples (Table 4.5).

Table 4.5: Yield of horticultural crops

Crop	*Yield (q)/ha*
Apricot (fresh)	20.00
Apricot (dry)	5.00
Apple	20.00

From Table 4.5, it is evident that the yield of fruit tree per unit area (ha) is satisfactory in view of the constraints prevailing in the region.

4.3.3. Vegetables

A large number of vegetables are grown in cold desert. The vegetables like, cabbage, cauliflower, knolkhol, lady's finger, onion, patato, tomato, carrot, spinach, turnip, radish, sugar beet etc. are grown in many parts of cold desert. These vegetables are consumed locally and the people learn a lot of money by selling these vegetables. Some seed production (vegetables) and multiplication centres have been developed in Kargil and Nubra valley. The production of Potato is the highest among the vegetables grown in the area (Table 4.6).

Table 4.6: Yield of vegetables

Crop	*Yield (q)/ha*
Leafy vegetables	25.00
Root vegetables	70.00
Cole crops	60.00
Potato	100.00

Source: Department of Horticulture, J & K State.

4.4. GROWING SEASON

Almost whole of the area is covered with snow clads for about 7 months and only 5 months are left where the activities can take place. Therefore, growing period is very limited. The seed sowing takes place in the month of April when the snow starts melting and the fields are cleaned of the snow creating the room for ploughing/sowing the fields. At this time, there is enough moisture in the soil. This provides a good opportunity for the operation of sowing of seeds. Therefore, all the physiological activities take place during this period. The short duration/span crops are cultivated. The vegetables are the ideal ones which get fit well in the frame-work of this period. However, the crops like wheat, grim, pulses etc. are also taken easily.

4.4.1. Cropping System

The most prevalent system of cropping is "Monoculture". Under this system, a single crop is cultivated through out the season due to its shortest life span (i.e. only 5 months or so). The cold desert owes essentially a single cropping season (i.e. from March/April to October). The data, elsewhere, revealed that net area sown is about 29.07 per cent and out of this, under double cropping system is just 10.66 per cent while rest of the bulk is single cropped area (i.e. 18.41%).

4.4.1.1. ***Why Monoculture :*** Whole area is the victim of heavy snow every year. The snow-fall starts from October and sometimes earlier to that and starts melting from March/April. The cereal crops have long duration and hence cannot be taken in the Rabi season. This leaves the chances for only the Kharif crops. These are cultivated in a single cropping season i.e. in the summer or Kharif which extends from March/April to October. The small millets China, trumba and Kangni) are seeded from April to August. Therefore, monoculture is the only alternate provision for cultivating cereal crops and vegetables.

4.4.2. Cropping Intensity

The gross cropped area varies from region to region in the State. Jammu region/province ranks highest in the gross cropped area. The gross cropped area in cold desert (Ladakh) is estimated to be almost negligible (Table 4.7).

Table 4.7: Gross Cropped area

Area	*Gross cropped area (%)*
Kashmir valley	40.20
Jammu province	58.98
Ladakh region	0.22

The cropping intensity, on the basis of gross and net cropped area, works out to be as 110.

4.4.3. Faulty Adjustment in Cropping Pattern

The success of agriculture mainly depends upon the sound cropping pattern. However, no system is supposed to be perfectly alright. But there are chances to improve the cropping systems in a particular area. The agriculture in Ladakh has several shortcomings (Chadha 1990). Following are the probable faulty adjustments in the cropping pattern:

4.4.3.1. ***Choice of species :*** This is the crucial factor in determining the soundness of the cropping system. There should always be good choice of tree/shrub species. But unfortunately, the cultivation of erosion-promoting crops leads to the disasterous condition. Not only this, but the cultivation with more than 30 per cent slope, ignoring even the soil conservation principles. Hence, the species should be chosen with every care according to the site conditions.

4.4.3.2. ***Dwindling forests***: In whole J & K State, the forests share about 21 per cent in the economy. The region is devoid of forest vegetation in comparison to other parts of the state. Whatever the forests we have that should be protected and preserved along with the increase through artificial regeneration. With the growing demands of the local people, there is high pressure on forests. Therefore, a progressive decrease in effective area under forest cover is visible. This is great setback to the region.

4.4.3.3. ***Degraded pastures***: The condition of the pastures is deplorable and the sources of forages are almost exhausted. Whatever, the pastures we have at present, they are being degraded. There seems no policy to declare the particular area as reserved for pastures or fodder trees (forages).

4.4.3.4. ***Wastelands***: A portion of land is regarded as westelands. These are the lands where no economic cultivation be practised. In actual sense, these lands are good for nothing. Now, efforts must be made to put these lands under viable use to harness at least minimum returns. It is now felt that every inch of land is crucial for the rapid growing population.

4.4.3.5. ***Faulty cropping system***: In the hilly tracts of cold desert, people now encroach the area. They generally plough those lands which are not easily accessible. They cut the mountains and make the strips and then plough it. This with the effect of snow, winds, run-off water etc. erodes away, leaving the area in bad shape and conditions. This practice should be discouraged in the interest of ecological stability.

4.5. ENVIRONMENTAL IMPLICATIONS

Agriculture occupies prominent place in cold desert. Naturally, the environment is influenced in a number of ways. Therefore, agriculture has some environmental implications some of them are as under:

4.5.1. The biotic environment has two major components viz. man and animals. It is generally advocated that most of the problems related to agriculture, forestry, environment etc. are man-made. There has always been a mismanagement of lands. The fertile lands may be put to grazing practices and vice versa. This trend/tendency leads to several environmental implications. There are several factors which create several types of environmental havocs.

Following are the factors worth mentioning:

(a) use of inferior seeds.

(b) substandard manures,

(c) improper irrigation facilities,

(d) indiscriminate use of pesticides,

(e) traditional methods of cultivation etc.

4.5.2. The domestic animals are the friends of inhabitants. The animals like goats, sheep, yaks etc. help in improving the condition of people and hence need careful attention. The animals supplement agriculture which supports the economic status of the inhabitants. The farms like poultry, pig, fish etc. should be given due weightage (Chadha 1990) for having a clean and sound environment.

4.5.3. The faulty management of pastures/grasslands, forests create many types of problems of agriculture. The soil and water conservation practices are not followed in true spirit which compels them to shift from one place to another. Due to agricultural practices, the top-soil is disturbed and eroded which gives rise to many a problems through which the environment is affected.

4.5.4. Now-a-days, the use of some insecticides, pesticides, non-traditional inputs is done on large scale. These ingredients do more harm to crops, fruits, vegetables, forage crops etc. They affect the environment in several ways.

4.5.5. The management of weather is of vital importance in maintaining the environment. Therefore, timely and proper forecasting about temperatures, rains, snow-storm, hail-storms etc. are essential for sound and healthy agriculture.

From the foregoing text, it is evident that agriculture affects the environment in a number of ways. Therefore, every care must be taken to make agriculture a viable media so that the standard of living of people may be raised along with the healthy environment. This will help in maintaining the proper ecological balance.

4.6. FARMING CONSTRAINTS

Agriculture is adopted in whole of the region. It has good opportunities but this profession faces a lot of difficulties. A few of the main constraints of farming are as follows:

4.6.1. Limited growing Season

As it is well-known fact that the area is covered with snow for a good period of time. The growing season is supposed to be of 5-6 months which is very less period. The growing period is from March/April to September/October. Thus, limited growing season is a major farming constraint.

4.6.2. Monoculture System

Mostly the system of cultivation is monoculture (18 % of the gross cropped area). Though double cropped area is also found but it is of too less extent. The seeds of different crops are sown during April/May and in some cases during August also (for example the vegetables). For agricultural crops at least 4/5 months are needed for maturity. Thereafter, no time is left for another crop (Rabi crop). Therefore, only Kharif crops are cultivated resulting in the monoculture system.

4.6.3. Diseases and Pests

The diseases and pests are major problems both for agriculture and horticulture crops. Most of apple production is severely attacked by scab diseases which takes a heavy toll of fruits. Also, a large number of pests attack the crops of agriculture and horticulture. Thus, every precaution should be taken to protect the crops.

4.6.4. Traditional Methods

In cold desert, the fruit growers are still following the traditional method of technology. They are still stick to some of the old techniques which include poor and old equipments and tolls, old methods of irrigation, cutting and plucking factors such as uneducated farmers, ignorant about the latest know-how. No serious efforts are made to introduce new equipment/tools. Similarly, there are old methods for agriculture. Therefore, traditional methods are the great hinderance in the way of both agriculture and horticulture.

4.6.5. Inadequate use of Fertilizers

Poverty is the main factor in using the adequate doses of fertilizers to agriculture (Dar 1986) as well as horticulture (the orchards). There is a great fear among the farmers that use of fertilizers will result in great loss of their crops. However, this is not based on the rational thinking. For example, in Kashmir Valley only 54.3 per cent orchards get fertilizers occasionally whereas the other 38.7 per cent do not get fertilizers at all. Therefore, same is the case with Ladakh as this region has almost been remained neglected. Further, no pesticides are used well in time.

4.6.6. Cost-heariness

Now, efforts are being made to use maximum non-traditional inputs and a lot of expenses incur on the purchase of good quality seeds. The inputs are too costly to buy them. But if, the same are brought then the out put is not upto the mark. This leads to the cost-heariness. As a whole the input cost is more than the output. It is true both for agriculture and horticulture (including the vegetables). Hence, there is urgent need to adopt new technologies to raise the production of different commodities (Anon. 1983)

4.6.7. No facility of Forecasting

It is found that the weather cycles in the region are unpredictable. Sometimes the falling of snow starts from September (too early) and sometimes it starts somewhat late. Similar is the case with melting of snow. The forecasting of weather or climate and the diseases can save a lot of wastage of fruits and agricultural commodities. The forecasting well in time can reduce the diseases and pests of horticulture, and agriculture. Now, it is the need of the hour that forecasting should be taken up by some institutions particularly in the region itself which will facilitate the work and will serve a great cause.

4.6.8. Slow Pace of Technology

Several new technologies have been developed in different parts. For example, CAZRI, Jodhpur has developed a lot of such ones. Therefore, on the same analogy these technologies can be tried in Ladakh. There are several agencies working in the area but unfortunately the slow pace of technologies is a great hinderance in the way of development.

4.7. FURTHER COURSE OF ACTION

Though whole of the region is hilly tract and the difficult terrains, there is a wide scope of adopting agriculture. This can be made sound profession. Same is the case with horticulture. Both these can be expanded to maximum extent if, some serious and true spirited efforts are undertaken. Following are some of the suggestion on the basis of which agriculture and horticulture can be flourished to harness the maximum benefits.

4.7.1. Integrated Approach

This is urgent required now because we cannot wait for the performance of single identity. It is always better to integrate all the possible disciplines to harvest their best and sweet fruits at a single time. Therefore, agriculture, animal husbandry, horticulture must be taken up side by side.

4.7.2. Improve the existing Cropping System

At present, mostly the old fashioned cropping system is followed in the cold desert. Now the time has come to change this trend with the growing demands of the population and the development of new technologies. For example, use of improved seeds, use of new equipments etc.

4.7.3. Adopt Silviculture

Silviculture is the art and science of cultivating forest trees/crops. There is lot of scope for this art on the naked hills and open spaces which are not used for other purposes. The steep sides of the mountains can be planted with the forest trees, fruit trees, fodder trees, fuelwood trees etc. Therefore, the northern sides can be afforested with these types of plantations. The flat lands can be put under vegetables etc.

4.7.4. Renovation of Lands

There is a great scope of introducing the forest trees/pasture grasses to those lands which are/were used for overgrazed practices. Similarly, the grass lands can be renovated easily. There is another provision for such lands that these lands must be closed for grazing so that the vegetative cover may come up well in time. This will prove best and strong way to utilize the unused lands.

4.7.5. Ley Farming

Ley farming can be adopted easily in such areas (Chadha 1990). In this system alternate rows of crops and grass strips are raised. This will give double advantages at one time. There are more chances of the success of such plans.

4.7.6. Researches

Research is the base of some sound programme as it gives the factual and tested truths. Through research, economy of the area can be increased. Following are the probable area where researches can be taken up on priority.

(a) research on fauna and flora, insect-pests of the area.

(b) soil testing programmes must be initiated,

(c) introduction of more medicinal and aromatic plants,

(d) identification of orchids, mushrooms etc.

(e) water and soil conservation, etc.

In conclusion, it can be said that agriculture is supposed to be the main-stay of the people. Though land holdings are very small, this profession can further be strengthened. The irrigation facilities through Government canals widens the scope of agriculture but still there is urgent need for some more work. The cropping pattern needs to be a drastic change in view of the fast pace of the time and development of new technology. Agriculture is supported by two other components viz. the horticulture and vegetables. They help to boost the economy of the region. Inspite of all these, some bottlenecks are there in the cropping system as well as in the landuse pattern, e.g. choice of species, dwindling forests, degraded pastures, wastelands etc. Agriculture influences environment in a number of ways. This faces a lot of difficulties. There are several constraints in the way of farming e.g. limited growing season, monoculture system, traditional methods of cultivation, inadequate use of non-traditional inputs, cost-heariness, slow pace of technology etc. For overall development of agriculture, some suggestions have been proposed. If, some serious efforts are made to implement some of the suggestions, then there is no doubt that agriculture will not flourish in the area inspite of some natural barriers.

"When all is said and done,
the final decisions about your land,
must come from you".

—*The Woodland Steward.*

Chapter 5

Animal Husbandry

"When a human-being or an animal eats a tree or a portion of the tree planted by a 'Momin', it shall always be considered a charity".

—*Hadith*

Cold desert (Ladakh) is situated at high altitude desert belt in the Indian sub-continent. It is held for a place of wild but uncompromising beauty. Though a very cold area, it gives a good scope for rearing livestock which is the essence of life of people. It is a source of milk and maintains other truth of life. An area of inhabitants without animals is as useless as a body without soul. Therefore, animal husbandry plays a vital role not only in the life of human but in ecology of the area also. Accordingly, the present attempt is in this direction to see the impact of animal husbandry on ecology of cold desert - the Ladakh.

5.1. ADVANTAGES

Animal husbandry is a main source of meeting several needs and demands of the locals. Many people depend upon them for earning their major part of income. Following are few advantages of animals (domestic ones) in Ladakh region.

(i) *Meat*: In the cold area, meat is more important than cereals. This gives more nutritions food which is considered very essential. Though the people are Buddhists, meat consumption is of high rate. Hence, animals give meet in abundant quantity.

(ii) *Milk*: Milk is taken in many ways viz. milk as such (raw milk), in the form of tea, by making Lassi (Butter milk), making Curd (Yogart), many types of sweets. The health experts are of the opinion that milk is the 'whole food'. It meets out all the nutritional requirements. It may be plain area, hot desert, high mountains, milk is very important for humans. In Ladakh, people are fond of taking butter-milk and butter-tea. This need can easily be meted out by animal husbandry.

(iii) *Wool* : It is the most economic commodity of the area. It is used in two ways viz.

(a) used for covering the body to protect in the winter.

(b) it is a good source of income (the people earn their livelihood).

Ladakhi wool maintains its identity and many products are made from it which are sold locally and exported also.

(iv) *Pashmina* : The pashmina sheep are found in cold desert which earns a good revenue. The people are expert in manufacturing the best pashmina shawls. These are liked by a large number of people. Therefore, pashmina industry can boost the economy of the locals.

(v) *Transportation* : In Ladakh, many animals are used for carrying the loads from the towns to the remote area. Many type of horses and donkeys are used for this purpose. Because of inaccessible areas, the use of animals is of utmost importance than the vehicles.

Thus, the animals are used in number of ways. They provide dung also which is supposed to be the vital factor in the productivity of soils. Similarly, animals can be harnessed in a number of ways.

5.2. CATTLE POPULATION

Most of the population in Ladakh lives in rural area. It has been reported that more than 80 per cent of population lives in village and only 12 per cent in urban (Sahni 1990). Due to small and scattered land holdings, the agricultural activities are performed manually. The farmers depend upon dzo (a male cross bred between cow and yak) for ploughing and transportation. Every farmer possesses one or two dzos. This indicates the standard of a family in the area.

According to Survey of 1981, there are following animals in cold desert (Table 5.1).

Table 5.1: Number of animals in Ladakh

Animal	*Number*
Changra (Pashmina goats)	1,40,000
Scrub goats	50,000
Angora Crosses	3,000
Changlook	9,000
Scrub sheep	60,000
Cross-Bred-marino	10,000
Cross-Bred-Cattle	4,000
Cattle (including Yak)	27,000

5.3. PRODUCTION OF KIDS AND WOOL

The indigenous livestock species are supposed to be inferior as they produce less milk and wool. There is a dire need to pay attention towards the proper development of the animals in the region. There are many yardsticks which require immediate attention. A few of them are as follows:

(a) a sound management,

(b) establishing fodder banks in the villages,

(c) provision of veterinary hospitals,

(d) establishing breeding farms,

(e) providing fertilizers on reduced rates,

(f) livestock experts for their proper advice,

(g) reseeding of the pastures, etc.

If, the above referred points are taken into account seriously, then healthy kids and more wool can be easily produced. It has been noticed that the cross-bred produce more milk than the local ones. The milk production (Table 5.2) has been compared with the local cattle and the cross-bred and it was found that the cross-bred cattle has upper edge over the local (traditional cattle).

Table 5.2: Milk production per day (kg)

Animal	*Milk Production*
Local Cattle	1.5
Cross-bred	5.0

The above data advocate that cross-bred cattle are more advantageous than the locals. Though the exotic species excel a large number of benefits, they pose some types of dangers, which are as follows, to the,

(a) plant species,

(b) grazing lands,

(c) domestic and wild animals,

(d) top soil and the pastures,

(e) water resources,

(f) elimination of some rare species of grasses such as Artemisia, Festvea etc.

Now, people (the experts) emphasize that local livestock should be improved for the prosperity of the people of Ladakh.

Different animals produce different amount of wool and pashmina. Table 5.3 shows the production of kids and wool.

Table 5.3: Production of kids and wool of different animals (Yearly basis)

Animal	*Number and type of kid*	*Wool (g)*
Pashmina goat	One kid	250 (pashmina)
Changlook	One lamb	2,000 (wool)
Scrub goat	One kid	500 (mohair)
Scrub Sheep	One lamb	1,000 (wool)

It is evident that the trend of producing the kids is the same. There are many more animal species which share a good proportion of the animals in the region.

5.4. ENDANGERED SPECIES

It is noticed that there are a large number of livestock species which are facing acute problem of extinction. This is a great threat to the ecology of cold desert. The cross bred species are creating danger for original/ indigenous and rare species. The Department of Animal Husbandry of J & K State is worried about several animal species such as Changlook, Yak, double humped camel, Zaskari horse, purik sheep, Appso, Hiniya sheep etc. It is urgently needed to save the local species of animals.

5.5. PROBLEMS OF FODDER PRODUCTION

Ladakh (cold desert) suffers from a chronic shortage of fodder which causes a severe constraint to Livestock Development Programme. The condition of livestock is in the worst shape due to inadequate fodder resources. However, some of the fodder species are cultivated by the farmers. The yield of some fodders is satisfactory which is depicted in Table 5.4.

Table 5.4: Yield of some fodder species

Crop	*Yield (q)/ha*
Alfalfa fodder (hay)	10.00
Alfalfa seed	0.12

Though the yield of few fodders is sufficient in the cold desert, there are a large number of constraints in the way of production. Following is the critical analysis of some of the constraints:

5.5.1. Climatic Diversities

The climate is the crucial factor in determining the performance of some identities/characters. The climate varies from extreme cold to extreme hot. Thus, there is a great diversity seen in the cold desert. It changes gradually and sequentially which accentuates the problems.

5.5.2. Mono-cropping

Due to high terrains and very short cropping season, mono-cropping is followed. Since, it is not possible to increase the area under forages, so there should be sincere efforts to make use of new technology for increasing the yield per unit area and per unit time. For this purpose, some forest trees which are excellent in fodder be encouraged to plant.

5.5.3. Degraded Alpine pastures

Still alpine pastures are found in pockets here and there. But unfortunately, the condition of these pastures is deteriorated due to some misuse by the dwellers. This is caused by the following factors:

(a) indiscriminate cutting practices,

(b) uncontrolled grazing,

(c) lack of non-traditional inputs (fertilisers),

(d) absence of leguminous components,

(e) overstocking in the area etc.

5.5.4. Seed Production

It is the major problem of fodder production in the area. If whole of the area is to be reseeded, a large quantity of seed will be required which incur lakhs of rupees. Hence, efforts should be earnestly made to produce the seed locally to reduce the expenses of seeds. For example, Alfalfa seed can be produced @ 0.12 q/ha in cold desert which is a good sign for prosperity of the fodder which can meet out the fodder demands to a great extent.

5.5.5. Weed Nuisance

Weeds are the greatest enemies of the pastures. Weed proliferation is a major constraint in pasture improvement. The weeds like, viburnum, typha etc. pose many hinderances in the way of pasture development. Weeds are the plants which can even grow in adverse and worst conditions. Therefore, the weeds take away the nutrients from the cultivated plants resulting a setback to the pastures.

5.5.6. Inadequate Breeding and Fertilization

Since, the cold desert has been neglected for a long time, so no good breeding programmes could be initiated with full vigour and strength. Therefore, a well unit breeding programme has to be started on war-footing. Moreover, identification of indigenous cultivates must be carried out side by side so that the great potential can be harvested with maximum use.

Extension education has been lacking in the area. The people still are ignorant about the use of fertilizers in the pastures/fodder producing areas. They should be encouraged for adopting new technology of cultivation of not only of agricultural crops but the pastures also. This will boost the production of fodder in the area upon which whole animal group depends. Shah (1986) suggested 80 kg DAP/ha for improvement of the pastures.

5.6. IMPACT OF ANIMAL HUSBANDRY ON ECOLOGY

Animals are also the essential component of environment. Same is the true with cold desert. After man, animals cause most extensive damage to plants (Negi 1983). As a whole, the animals which cause damage to forests may be categorized into two groups, viz.

(a) Domestic animals.

(b) Wild animals.

It has been opined that the injury/damage is caused more severely by the domestic animals.

5.6.1. Feeding Habit

This is the vital character which determine the nature of animals which affect the forest, the environment and ultimately the ecology of the area. The domestic animals can be divided into two groups depending upon the feeding habit.

5.6.1.1. ***Graziers:*** These are the animals which feed on grasses and herbs. Examples can be horses etc.

5.6.1.2. ***Browsers:*** These are the domestic animals/others which feed on the twigs, shoots etc. The examples are goats, sheep, double humped camels etc.

5.6.2. Damages

Both types of animals i.e. the graziers and the browsers cause heavy damage to forest and other vegetation. Some of the visible and invisible damages are as follows:

5.6.2.1. ***Regeneration problem:*** It is a common practice in the area that animals are grazed in the open fields, on slopes, mountains etc. They graze the desired species of plants alongwith the grasses and herbs. This practice leads to the creation of regeneration problem. Thus, it affects the coming up vegetation and composition of the crop. This creates the environmental hazards.

5.6.2.2. ***Erosion problem:*** The cattle graze in the field and make the soil compact with the repeated effect of hooves. There may be a considerable change in water absorption capacity of soil. The soil also gets loosen which causes a heavy run-off. This will lead to the erosion problem as the soil is eroded heavily.

5.6.2.3. ***Death of seedlings:*** The over/uncontrolled grazing leads to the exposure of roots of the plants. Not only this but the severity of the seedlings also takes place. The trampling action of animals leads to the ultimate death of growing seedlings in the grazing prone areas.

5.6.2.4. ***Soil problems:*** There are indications that many problems of soil are created simply by the action of indiscriminate and heavy/over grazing. These actions cause many problems related to soils. A few of them are as follows:

5.6.2.4.1. *Compactness:* The heavy grazing leads to compactness of soil which is a hinderance in the way of coming up crops. This affects the local environment greatly.

5.6.2.4.2. *Porosity and Aeration :* The porosity and aeration of the soil is heavily affected by the action of animals grazing. These will lead to many types of havocs.

5.6.2.4.3. *Others:* Soil is affected in many other ways also such as,

(a) increase in surface run-off,

(b) accelerated erosion,

(c) decrease in water absorption and retention capacity,

(d) leads to crumby structure of soil.

Thus, soil is affected in a number of ways by the action of animals particularly the domestic ones. These effects influence the environment and ecology of the region.

In conclusion it can be said that animals are the essential component of environment. Ladakh (cold desert) though plunged into a number of problems, gives good scope for boosting up animal husbandry. It meets out many demands of the people. It excels a large number of advantages. Many species are found which are at the edge of extinction. Fodder is in short supply in the region. The animals have many kinds of impacts on ecology. There is urgent need to improve the animals as well as the crops. The domestic animals cause several types of damages which affect the ecology and environment.

Chapter 6

Forest and Forestry

"Food is subservient to Forests"

—*S. Noor-u-Din Wali*

6.1 THE FORESTS

"Poems are made by Fools
Like me,
But only God can make
A tree."

—*Joy Cee Kelmer*

The forests are the Green Gold of any nation. They are regarded as the great asset. They are not only of paramount importance to the State because of being a rich source of revenue but also for maintaining the ecological balance. The development processes and the scientific evolutions have reversed the relationship of human that to the environment. Hence, such attitudes have degraded the forests and natural resources. This has led to the human race almost at the threshold of extinction. Although the forests are worth the name only in cold desert, there is an urgent need to protect and develop this indispensable possession. Accordingly, the present attempt is in the direction to view the position and role of forests and forestry in the cold desert. The forest are useful to society in a number of ways.

6.1.1. Benefits of Forests

In some of the cases, Wood Consumption is considered as a 'Yardstick' of measuring prosperity of a nation. The forests are the soul of the poor as he is dependent upon forests in a number of ways. The benefits are judged in the following 2 ways (Negi 1983).

6.1.1.1. *Direct Economic Benefits*

(a) source of energy;

(b) best source of employment;

(c) small scale industries;

(d) provide individual demands;

(e) source of raw material;

(f) good source of revenue;

(g) tourism development;

(h) commercial timber; etc.

6.1.1.2. *Indirect Economic Benefits*

These are also called intangible benefits of forests. A few of them are as follows:

(a) Ameliorating Climate:

  (i) Temperature (high and low);

  (ii) Rainfall;

  (iii) Humidity;

  (iv) wind

(b) Moisture Conservation;

(c) Flood Control;

(d) Aesthetic role;

(e) Pollution Control;

(f) Soil and Water Conservation.

6.1.2. Extent of Forests

The cold desert is almost devoid of any significant vegetation. However, there is some amount of vegetation including the forests. The extent of forests is viewed as follows:

6.1.2.1. ***Landuse pattern*** : For evolving any viable programme/system in cold desert (Ladakh), the knowledge of existing landuse pattern is of paramount importance. The barren and uncultivable land constituted as much as 44.96 per cent and another 13.95 per cent area accounts for cultivable waste land, out of the total land. However, net sown area comprised of only 29.07 per cent whereas out of this only 10.66 per cent area is sown more than once. In Ladakh, the land under miscellaneous tree crops and groves accounted for only 4.65 per cent (Table 6.1).

Table 6.1: Landuse pattern in Ladakh

Particulars	*Area (ha)*	*Per cent of total area*
Total area according to village Papers	64,500.00	-
Forests*	460.00	0.71
Land under non-agricultural use	3,000.00	4.65
Barren and uncultivable land	29,000.00	44.96
Permanent Pastures and grazing lands	1,215.00	1.72
Land under miscellaneous tree crops and groves	3,000.00	4.65
Cultivable Wasteland	9,000.00	13.95
Fallow and Current fallow land	73.00	0.11
Net area sown (i.e. area sown more than once)	18,752.00 (2,000.00)	29.07 (10.65)

Plantations and other undemarcated forests are not included which roughly come out to be 1,000 ha.

It is evident from Table 6.1 that area under forest and permanent pastures (including that of grazing land) seems to be deplorable which is estimated as 0.71 per cent and 1.72 per cent respectively.

6.1.2.2. ***Biological spectrum*** : A gradual transition of vegetation from desert type into temperate is observed where the southern boundaries of this region touch Kashmir and Himachal Pradesh. Similar vegetational changes from temperate to xeric type occur as one enters into the regimes of the region through Zojilla (La-Pass). The herbaceous elements are essentially similar to those found in Kashmir Valley, but the tree elements disappear all of a sudden. The open sward generally consists of *Lolium, Bromus, Sibbaldia, Arenceria, Taraxacum, Plantago* etc. However, *Juniperus recurva* forms the major shrub elements while the tall shrubby vegetation is dominated by *Pranges pabularia*. Kargil is semi-arid with higher precipitation than Leh. The vegetational cover does not exceed 50 per cent in unirrigated area (Misri 1980). Moist soils on the banks of the rivers and streams may support conspicuous vegetation consisting of *Artemisia, Rosa, Corydalis, Euphorbia, Heracleum, Calamagrostis* etc.

About 662 plant species are found in different matches of cold desert. These are spread through 51 families with maximum representative species in *compositae, Cruciferae, Labiatae, Polygonauae* and *Gramineae*. Following *Raunkier* (1934), the species can be grouped in the biological spectrum (Table 6.2).

Table 6.2: Biological spectrum of Ladakh region

*Per cent Number/ Life-form Class**	*Spectrum for Ladakh*	*Normal Spectrum*
Phanerophytes (PH)	2.62	28
Nanophanerophytes (N)	4.84	15
Chaemephytes (Ch)	26.83	9
Hemicrytophytes (H)	9.61	26
Geophytes (G)	21.31	4
Hydrophytes (HH)	0.62	2
Therophytes (TH)	33.18	13
Lianas (L)	0.00	0
Epiphytes (E)	0.00	3
Parasites (P)	0.09	0

* from Raunkier (1934).

6.1.2.3. ***Forest type***: Champion and Seth (1968) have classified the forests of India in various types and sub-types. In the broad classification of the forests of Jammu and Kashmir State, the forests of cold desert (Ladakh) have been classified as 'Alpine Forests Type'. These forests are found well above the temperate zones. A large number of species are met with in the region. Some of the common species, found in the zone are, High level Fir *(Abies pindrow)*, Kail *(Pinus wallichiana)*, Junipers, *Quercus semicarpifolia, Populus ciliata, Betula bhojpatra, Salix* spp., *Rhododendron arboreum*, and a variety of grasses and wild flowers.

6.1.2.4. ***Other plants/trees of interest*** : The entire area from Zojilla to Chushul has rightly been classified as 'Alpine Scrub Zone' because very few species of trees are found and that too grow only in some specific pockets covering almost an area of about 460 ha. The tree species are,

Salix dephnoides, *S. alba*, *Juglans regia*, *Populus deltoides*, *Morus alba* etc. Among woody shrubs, however, are, *Myricaria elegans*, and some wild roses. Ladakh is a rich store-house of medicinal plants. *Physochlaina preleata* is cultivated in fields. *Ephedra gerardiana* is met with on very dry rocky areas and similarly Artemesia spp., is found throughout the tract. *Podophyllum emodi*, *Onosma* spp., and the *Saussuria lappa* are available in Drass, Kargil, and Zanskar areas.

The availability of fuelwood is an acute problem. This is probably due to the following reasons, (a) Very limited and inadequate forest resources of the region, and (b) higher per capita consumption due to extremely cold and prolonged winter. The estimated annual consumption for both Kargil and Leh may be more than one lakh metric tonne. About 1/3 of this demand is meted out with local shrubs including that of *Artemesia* roots, stacks and the tree vegetation whereas 2/3 of the requirements are met with dried cattle dung (which otherwise is a precious commodity to increase the fertility of the soil).

6.2. FORESTRY

True Forestry is Statesmanship of the finest type.

—F.C. Ford Robertson

Forestry in one or the other form has been the salient feature of the cold desert. Nature has also been kind enough in blessing the area with some of beautiful forest pockets. Equally, man has also been active from the very beginning in protecting and producing the forests. The efforts are still on to add something to the region. People are the followers of Lord Budha and the Mahatma advocated the idea to,

"Everyone should plant a tree in five years".

—Lord Budha

This was the need at that time which is now many times fold. It is due to the tremendous increase in population with whose dimension, the demands have also increased in the same proportion.

6.2.1. Traditional Practices

There is very little forestry in the existing cultivable land. However, practice of raising trees in and around agricultural crop land seems to be vogue partially, probably as an insurance against high wind velocity. Due to the acute fodder shortage, the farmers collect green leaves and tender twigs of such trees to feed the cattle when the area is covered with snow. There are some practices which are followed in the cold desert which need to be mentioned.

(i) The trees, mostly poplars, willows, are planted along the river beds in 2-3 rows. The cultivable land is then earmarked on the opposite side of the tree plantations and cultivation of wheat, grime and some pulses is carried out.

(ii) The trees of *Salix* spp. are planted around the farm on the ridges in close proximity to each other.

(iii) The poplars and willows, are planted, however, with no definite ratio, around the farms and also sparsely distributed in crop land.

(iv) The trees are planted only in the crop land and around it, a ridge or wall made of mud (Kacha wall) and stone measuring, generally 2-3 ft, is erected.

6.2.2. Scope of Forestry

Forestry in the form of social forestry is meant for service of society at large. According to National Commission on Agriculture (1976), the components of social forestry and their probable role in the development of cold desert can be visualised as,

6.2.2.1. ***Agroforestry*** : The main aim of this forestry is to meet the requirements of human as well as cattle population. The immediate need of the farmers is to have 5 Fs viz. food, fuel, fodder, fertilizer, and the fibre. The region has difficult terrains but due to its vastness, there is ample scope to popularise agroforestry. In the light of the above arguments, following probable systems (King 1978) could be tried:

(a) Agri-silviculture (Agriculture crops + Tree components).

(b) Sylvo-pastoral (Tree component + pastures i.e. the forage crops).

(c) Agro-sylvo-pastoral (Agriculture crop + Tree component + grasses/forages).

(d) Agri-horti-silviculture (Agriculture crop + Horticulture + Tree component).

The major contribution of agroforestry could be visualised through the following needs:

6.2.2.1.1. *Fodder Development*: Ladakh is famous for the production of valuable 'Pashmina'. But this industry could not go ahead due to limited scope of rearing of Pashmina goats in the scarcity of fodder resources. Therefore, similar type of setback is to the development of wool industry and dairy farming. In promoting these industries, production of fodder and grasses is urgently needed and agroforestry can meet these challenges to a great extent.

6.2.2.1.2. *Energy plantations* : Due to negligible existence of natural forests, there is no means of getting fuelwood which is the basic need of the inhabitants. Agroforestry can pave the way for raising energy plantations by selecting suitable species depending upon the adaptability of a particular tree species. It will widen the scope of harnessing the energy for fuel purposes and will solve the problem to a satisfactory extent.

6.2.2.2. *Extension forestry*: The inhospitable climatic and edaphic conditions prevail in cold desert. Consequently, tree planting is a problem. However, extension forestry programs can be taken up in the following ways:

6.2.2.2.1. *Raising Plantations*: At present, timber in the region is brought from Kashmir Valley at the exorbitant rates which are even beyond the capacity of people. Hence, raising plantations with suitable and fast growing tree species is urgently needed. In addition to naturalised species, artificial afforestation should be taken up. Mathur and Sharma (1983) suggested to introduce some species of Eucalyptus.

6.2.2.2.2. *Erecting Windbreaks*: The cold desert experiences high wind velocity. Hence, windbreaks can play a vital role in saving the other crops. They have soothing effect on micro-climate. Pathak (1958) recorded beneficial effects of windbreaks of *Prosopis cineraria* in Rajasthan desert. Similar strategy can be developed easily in cold desert. This, however, depends upon the suitability of species.

6.2.2.3. *Afforesting degraded lands* : This type of land is found in abundance. They provide ample scope for afforestation programmes. This will certainly enhance the forest wealth in the region. This can be viewed as follows:

6.2.2.3.1. *Soil Conservation*: Since there is practically no rainfall, so water erosion is not a problem except for bank cuttings by the rivers and nullas. But the absence of vegetation cover and extreme type of climate invite the wind erosion which is a major nuisance in cold desert. The magnitude of this problem is too big and anti-wind erosion measures are costly. The only alternative is growing of forest trees which are suitable for soil conservation.

6.2.2.3.2. *Efficient Landuse*: The barren and culturable waste lands account for more than 61 per cent which is a setback to proper landuse planning in cold desert. If, sincere efforts are made in putting these lands under proper use, this can help in boosting the economy. If class V and VI waste lands are planted with some suitable grass species and trees, then a judicious use of such lands can be made. The trials should be laid out with regard to local flora.

6.2.2.4. ***Recreational forestry*** : Ladakh has a great potential and scope for such type of activities. The following components of recreational forestry can be viewed:

6.2.2.4.1. *Amenity Plantations* : For an overall development, a nation needs wealth in the form of material goods, but a beautiful environment is just important. The locals are the beneficiaries of such plantations. Therefore, to harness maximum benefits approach is very essential to seek their cooperation (Sagwal 1985).

6.2.2.4.2. *Wildlife*: Wildlife is an integral part of forests. Both are complimentary to each other. Vegetation is, however, scanty but the variety of flora is capable of sustaining a wonderful and magnificent wildlife in the region. Therefore, forestry provides all necessities for growth, protection etc. to wildlife.

6.2.2.4.3. *Recreation Resorts*: Land utilization aims at the exploitation of resources of lands for the welfare of mankind. People after hardwork of daily life want to relax for a while and hence they need recreation resorts. They can enjoy at picnic-sites, game-resorts, cafes etc, which can be established in the forests or nearby the forests. There is an urgent need to develop recreational forestry in cold desert.

6.2.3. People's Participation

People's participation is the the need of the hour. It is questionable why the foresters fail to involve the people in management of forests? There are probably two reasons, viz. first, though the people are integral part of forest management does not occur in the traditional values. The share of locals and the tribals did not arise, and, secondly, the growth of the forests as a resource base for supply of wood commodity does not fit in the framework of increasing demands - supply battle. Hence, sincere efforts are needed to involve the local people in forestry projects. If, possible the bottom up policy should be adopted.

6.2.3.1. ***Scope of participation***: Forests are the main stay of the hill dwellers. There are many areas where the participation of the people is a crucial factor (Shrivastava 1989) are,

(a) the protection of forests from illicit felling,

(b) efficient management of productive forests,

(c) control of grazing by providing regulated grazing and avoid overgrazing,

(d) prevent the encroachments,

(e) prevention of fires,

(f) rehabilitation of degraded lands and forests,

(g) afforestation of wastelands,

(h) soil and water conservation,

(i) conservation of wildlife.

6.2.3.1.1. *Government Forests* : The government Forests face several types of problems, viz.

(a) theft of timber,

(b) grazing,

(c) fire, and

(d) law enforcing agency - forest staff.

It is, however, necessary to provide sufficient incentives before taking any approach towards ensuring the participation of the locals. Therefore, following suggestions would be appropriate to take,

(i) *Incentives* : There should be some provision to tackle the problem of theft. Assistance should be taken from the villagers in nabbing the theft of timber, fuelwood etc. It should be given to the staff of forest as reward and must be distributed among the villagers.

(ii) *Provision of subsidiary*: The commodities like timber, fuelwood, fodder etc. should be supplied to villagers at subsidised rates. This will encourage the people to participate in the forestry projects.

(iii) *Local Committee*: There is an urgent need to protect the forests. If local elements take care of such assets, then it will be more effective. Regular patrolling, controlling grazing, prevention of fires etc. should be taken on priority.

6.2.3.1.2. *Outside Government Forests* : This types of forests/plantations must be saved other than that of forest of Government. This can be taken into two ways, viz.,

(a) Social Forestry.

(b) Afforestation.

6.2.3.1.3. *Role of Women* : There are a large number objectives to be achieved with regard to forestry. There objectives may be,

(a) increase in the production of fuel, fodder, and the fruits,

(b) creation of employment opportunities,

(c) promoting environmental conservation etc.

These objectives, however, can be achieved only with the help of women who are the primary consumers of forest produce. Hence, about 50 per cent of the extension workers should be women folk to make forestry a successful programme.

In conclusion, it can be said that forests are the great assets of a nation. There is urgent need to protect the existing forests and for enhancing, we should go for afforestation programme for maximum area which are accessible. The forests serve the people in a number of ways. Direct and indirect benefits of forest can be visualized. Forestry plays a vital role in the life of the people and which helps in maintaining the ecological balance. A need of people's participation is felt urgently so that the forestry can be made a viable programme. This will add to the overall development of the people and that to the region.

Chapter 7

Environmental Pollution

"Treat nature well,
And nature will treat you well;
Hurt or destroy nature,
Nature will soon destroy you".

—Aldous Huxley

Environmental pollution is a problems created by man himself. This is because he is upsetting the environment where he lives. Elsewhere, Alexander Pope, made an observation that 'proper study of mankind is man.' This is only possible when one studies the environment. The pollution is an ecological problem and thus demands an equal understanding of both the environment and the organisms. The problem of pollution has been causing concern to the living things and their environment. Now, man is too eager to get maximum benefit in terms of production (i.e. the crop production) and use fertilizers which create many types of problems. Cold desert is also affected by this nuisance problem. Though the extent and magnitude of the problem is difficult to estimate, there is an urgent need to know the agencies responsible for environmental pollution. The present attempt is in this direction to explore the possibilities of the causes and preventive measures of environmental pollution.

7.1. DEFINITIONS

Before dealing with present topic, it is utmost important to know some of the relevant definitions. Following are the definitions which encounter in the text.

7.1.1. Environment

It is the sum total of living conditions and the parameters setting up the relationship between the living and non-living things.

7.1.2. Pollution

Any substance added to the environment as a result of human activities which has a measurable and deterimental effect on the environment.

The problem of environmental pollution is becoming very acute day by day. The term pollution is so frequently used that the changes are in confrontation with other things.

7.2. TYPES OF POLLUTION

Though the estimation of the extent of pollution in the region is extremely difficult, if, not impossible, there are mainly two types of pollutions met with. Following are the two types:

7.2.1. Water Pollution

The sedimentation due to water erosion is by far the prominent cause of water pollution. The Indus river, running throughout the region, is getting polluted by excreta which poses as great threat to the inhabitants. The people drink water of this river which is a direct invitation to several types of diseases and the disorders of the human body. Thus, highly polluted water accumulated in the river and other sources is the major cause of cholera, typhoid and mainly the dysentery. The siltation rate reported ranged from 8 to 68 t $acre^{-1}yr^{1}$ (Kangoo 1989). However, this depends upon the landuse in the area.

7.2.2. Visual Pollution

Cold desert is situated in the inner parts of Himalaya and possesses a peculiar type of (i) ecology, and (ii) environment. This part is worst affected by visual pollution which brings several types of havocs. The wastes in the form of solids, are found lying here and there which constitute the garbage pollution. However, all the *three* parts of Jammu and Kashmir State are suffering from visual pollution. The rubbish heaps are the store-houses of several diseases. The cattle grazing over such garbage dumps yield inferior quality milk. This pollution is creating grave situation and the government is much concerned with this problem. However, it can be tackled to a great extent by afforestation programme on large scale. This will add greenery to the area and thus helps the landscape in maintaining ecological balance.

7.3. FACTORS AFFECTING POLLUTION

Above all, the man is responsible for creating all types of environmental pollution. This holds good in case of cold desert also. There are several factors which affect pollution. Following are the probable factors:

7.3.1. Population Explosion

Though cold desert owes less population in compares to other parts of J & K State, there is an increasing trend in Ladakh province. According to 1981 Census, the population density was 2 people/km^2, but now, the figure has gone even upto 5 people/km^2 in many parts of the region. The increasing population is posing a severe threat to the available resources. More population means more problem of environmental pollution. The increasing population is of great concern as the land (a constant factor) cannot be expanded. Thus, the increased population is a direct and unbearable burden on this valuable asset (land).

7.3.2. Reckless Felling

Removing vegetation in any form leads to the naked condition. Thus, devoid of vegetation creates many problems. Reckless felling creates such situations where the changed climatic condition poses a severe threatening to ecosystem. Most of felling is done for constructing industrial and housing complexes. This is due to the fact that everybody desires to live in peace and this abode of peace is destroyed by cruel jaws of greedness of man. If, this continues, the day is not far away when the whole area will be turned into barren tracts. This condition may compel to lead to the havoc like situation of environment.

7.3.3. Tourism

Tourism is the mainstay of the economy of the state of Jammu and Kashmir, and hence for cold desert also. The forests, whatsoever are there in the region are being cleared for creating structures of complexes and the colonies. These are meant for tourists who visit this above of peace. This leads to shrinkage of forest wealth. The wastes and the night soils (e.g. in Indus Valley particularly along the bank of the river) are being added to the water resulting in polluted form. The landscape of cold desert is deteriorating day-by-day and becoming ugly with a consequent of loosing its purity and virginity which is a setback to the ecology. The indiscriminate interference of human has already changed the climatic pattern and dried up natural springs (Chadha 1990). This has a great impact on the environment.

7.3.4. Cattle Grazing

This is the worst and nuisance problem prevailing in the region. The uncontrolled cattle population and various types of grazings are the prominent causes of land degradation and deforestation. There are many ill-effects of grazing particularly the over-grazing viz.,

(a) the forest/open floor gets hardened which results in reduced percolation of rain-water and more run-off.

(b) the soil erosion takes place. The water table, which is otherwise very crucial factor, goes down.

(c) severe damages to the emerging/growing seedlings. Also, this will pose a great hinderance in regeneration (of course both types i.e. natural and artificial).

(d) with the increase of cattle population the yield of grass which is an essential commodity, will reduce to a great extent.

All these effects generally disturb the natural ecosystem. Hence, to control/regulate the grazing is of vital importance in the light of its ill-effects.

7.4 ENVIRONMENTAL PROBLEMS

Cold desert is implicated with several environmental problems. Following are a few problems which are viewed as.

7.4.1. Soil Erosion Problem

Ladakh being a cold desert experiences extreme cold, dry and arid type of climate. This is infringed with the problem of soil erosion. There are several agencies which prompt the erosion, viz.,

(i) improper land use,

(ii) overgrazing,

(iii) deforestation,

(iv) conventional method of agriculture,

(v) unplanned constructional work etc.

The debris accumulates at several places which proves to be hinderance in the way of flushing water during summer and results into floods and this removes the soil ruthlessly causing an acute problem of soil erosion.

7.4.2. Wildlife Conservation

Due to non-availability of required sources for survival of wildlife, these remains always a danger for extinction of some of the species of birds and mammals. This is attributed mainly due to two main reasons, viz.,

(i) hostile environment, and

(ii) non-availability of good and sufficient fodder for these animals.

The vegetation is of xerophytic type and scanty in quantity and the mammals donot like such vegetation. The birds generally migrate to lower elevations during severe cold. Therefore, there is urgent need to conserve the wildlife which is the integral part of the environment. Some of the measures should be adopted for protection of wildlife. There may be the following relevant points which need immediate attention.

(a) create fooder banks,

(b) cull out the dry stock of animals,

(c) establish veterinary hospitals to cover the maximum area,

(d) provide proper and timely vaccination,

(e) creating sanctuaries in the area.

7.4.3. Impact of Tourism

Tourism is an attracting feature of cold desert (Ladakh). A lot of facilities have been created by the authority for hosting the tourists not only from India but abroad also. The region of Ladakh is superb with the following attractions:

(a) Alpine pastures,

(b) Hot springs,

(c) Green meadows,

(d) Cascades and Waterfalls,

(e) Rainbows,

(f) Glens,

(g) Valleys,

(h) Excellent fauna and flora etc.

Inspite of all these, hostile climatic set up has posed hurdles in the ways of its progress. Even then, it is famous for its cultural heritage and virginity. This is supposed to be the essence of life. Due to this identity, tourists are attracted from different parts of the World. Now, it is an urgent need to save this beautiful and precious baunty. Every effort should be made to maintain the balanced ecosystem. For this purpose, following steps must be taken into consideration:

(a) make arrangements for boarding and lodging,

(b) save the tourist area from visual pollution,

(c) more cultural programmes should be shown to the visitors to maintain their valuable heritage,

(d) raise economic level of the dwellers for increasing their standard of living.

7.4.4. Saving Water Resources

Water seems to be a scarce commodity in the region. The rainfall is scanty as only 20 cm annual precipitation is received and that too during the months of January, March, and August. For maintaining ecological balance, artificial sources of water must be created. Therefore, a judicious and proper utilization of water resources should be made to change ill-fated conditions of poor-resource dwellers. It is a well known fact that water is the essence of life of people. No life is even possible without water. It is the need of the day to harness maximum number of sources of water.

7.4.5. Fire Hazards

Fires are the features of summer season irrespective of type of vegetation. They take a heavy toll of forest and other vegetation and reduce the asset into mere ashes within no time. The uncontrolled fires may spread to the nearby areas (forests). They will hamper regeneration and pave the way to accelerated erosion and destroying the landscape.

7.5. POLLUTION CONTROL

Pollution is on advancing pace with rapid increase in population. Its control is urgently needed to maintain our environment clean. National Botanical Research Institute, situated at Lucknow (U.P.) is the only prominent institution in India which deals with research on various types of plants. Now, pollution research is also being dealt with the this institute to cater the need of the country. Following are the proposed directions where research work about pollution is being carried out:

(1) injuries to which the plants are victim and subsequent report preparation thereof.

(2) effect of pollution on the external and internal parts of plants and kind of pollution.

(3) to find out species which are resistant to pollution and grading them for their suitability to polluted areas.

7.5.1. Control Measures

Efforts should be made to combat the evil of environmental pollution. Though there is no guaranatee to control pollution completely, it can, however, be reduced to a great extent. Sagwal (1991) has suggested the following:

7.5.1.1. Ceasing unscientific exploitation of natural resources. Exploitation is taken in two senses viz. positive and negative. There is no harm even to exploit the natural resources in negative sense but it should not be in unscientific manner. For example, harnessing the water for useful work, it should be stored in well built reservoirs so that maximum benefit can be taken.

7.5.1.2. ***Increase vegetative cover***: It is urgently required now to increase vegetative cover so that a good tuft can be obtained. This will help in protecting the floor from the onslaught of any type of erosion viz. water/air erosion. The plantation works should be carried out throughout the region on war-footing. Very fast growing tree species like Poplars, Willows etc.; can play a vital role in increasing the vegetative cover.

7.5.1.3. ***Saving existing forests***: There is now need for planting as many trees as possible. This will fill the gap which is created by illicit fellings. The protection of existing trees should be taken on priority. This will help in maintaining ecological balance.

7.5.1.4. ***Regulate grazing practices***: Over-grazing is an ill-fated practice which leads to the devastation. Therefore, more research is required to exploit this in a proper way. Grazing should be regulated at any cost.

7.5.1.5. ***Sound planning of tourism***: The tourism is a flourishing industry in Ladakh. Hence, framing of a sound planning of tourism is urgently required. Also, stop the encroachment in the areas of forests, lakes, rivers, and catchment areas. This will save the ecosystem of cold desert.

7.5.1.6. ***Stop the garbaging*:** The garbages lie here and there in cold desert area which cause several problems. They create problems to destabilise the ecosystem. Hence, there should be a sound planning of tourism and there should be arrangement to dispose of the garbages as soon as possible.

7.5.1.7. ***Sanitation of the area is essential*:** There should be a provision to clean the area and water particularly of the adjoining region of the river Indus and other catchments.

7.5.1.8. ***Controlling population*:** Ladakh still has the lower population records in comparison to many other regions/states. Even then, with the increase of population, cold desert is also worst affected. The pollution can be reduced to a satisfactory extent by controlling human as well as livestock population.

It is concluded that environmental problem is very serious as it affects the whole living system in the area. Here, two types of pollutions are found viz. water and visual pollution. A large number of agencies, e.g. illicit/reckless felling, population explosion, tourism, cattle grazing etc. are responsible for environmental pollution. There are several environmental problems which occur in Ladakh. Some suggestions for controlling the pollution have been given. If, sincere efforts are made, then pollution can be tackled to a great extent, if not fully.

Chapter 8

Wildlife

"Wildlife! That is how we refer to our magnificent animals and birds that brighten our lives. Inspite of our culture and civilization, in many ways, man continues to be not only wild but more dangerous than any of the so-called wild animals. Life would become dull and colourless, if, we did not have these beautiful animals, and birds to look at and play with. We should, therefore, encourage as many sanctuaries as possible for the preservation of what yet remains of our wildlife".

—*Jawahar Lal Nehru*

Wildlife is a science that deals with all free-ranging vertebrates in their naturally associated environments. The objectives of wildlife management programmes may be to control or to create congenial favour to the abundance/distribution of the vertebrate species. However, it must be kept in mind that free-ranging vertebrates must be unfenced or atleast should be kept in very large enclosure. Wildlife management is the art of making land to produce valuable population of wildlife (Bailey 1984). Wildlife is a diverse resource of the environment.

The forests form an integral part of environment and signifies about the greatest asset of our country, particularly the cold desert. They add considerably to the states revenues and forest products including that of the wildlife to our export earnings. However, the local people are not benefited in the real sense. Ladakh, the cold desert of Jammu and Kashmir State, abounds a variety of wildlife. Inspite of the most different terrains in the region, it is blessed with a rich fauna which is a matter of great pride not only for the region but for whole of the country also. Here, in this Chapter, an attempt has been made keenly to explore the possibilities of wildlife in addition to the management and conservation of existing species.

8.1. SALIENT FEATURES

The region has a typical type of ecosystem. Wildlife has peculiar and similar type of characters in the cold desert. Mostly, the wild animals exhibit a number of common features. A few of them are as under:

(a) they oftenly inhabit the high grounds,

(b) have high powerful ad strong sight,

(c) have very/high power of smell and hearing,

(d) they have thick hairy coats on the body like a woollen puffed,

(e) the males have usually bigger horns,

(f) the individuals of different species move about in herds, and

(g) can tolerate the extremes of cold.

8.2. MAJOR SPECIES

The region is having abundant wildlife. However, in the present text, only a few prominent species have been taken into consideration.

8.2.1. Bos grunniens (Wild Yak)

Locally, it is known as 'Dong'. This is a wildlife species of giant nature possessing a height of about 1.75 m at the shoulder. The body of yak is covered with black tufted hair which go down to the ground giving a clumsy appearance. The feeding habits of yak are very simple which include masses, lichens and grass shoots. It throves very well above 4,600 m elevation.

8.2.2. Carpra falconeri (Markhor)

This is one of the finest animals of the region of Ladakh. In real sense, this species belongs to the family of goat and recognised as the 'King' of wild goats. The animal is about one metre in height at shoulder point. The males have very fine black beards and colour of body hair is usually light yellowish/reddish brown. The animals roam only in the morning and evening. The male and female live aloof with an exception to the mating season. This species is found only in Kargil area.

8.2.3. Equushominus kiang (Kiang)

The animal is very handsome, brown from the above side and white below the belly. The animal measures about 1.25 m at shoulder. It has black mane. The animal is midway between the ass and mule and found throughout the region.

8.2.4. Felix uncia (Snow Leopard)

This is a very famous animal and is locally known as 'Shun'. The total length of the animal is about 2 m and measures about 1/2 m in height. The colour of the body is grey/white with a creamy tinge. This preys upon some of the local wild animals. It is nuisance to goats and sheep during winter season because it comes down from the high reaches to the lower/ foot hills.

8.2.5. Gazella picticaudata (Goa)

The other associate of goa is known as chiru (local name). These are two handsome animals, *G. picticaudata* has very beautiful horns and measures at more than 1/2 m in height at shoulder. The body is sandy-grey in colour. The animal is covered with a coat of soft-fur. Mostly, the goats are inhabited in chang-Chenmo and Rupsu areas.

8.2.6. Ovis nathura (Bharal)

Ovis nathura is known by the name, 'Blue sheep' and locally in Ladakh, it is called as 'Napo'. The sheep is, however, handsome (male) and beautiful (female). These are the sturdy animals which can tolerate the cold to a great extent. The body of the animal is generally thick grey coated which is bluish on out side and white from beneath. The horns are of smooth with olive-green colour. The sheep ascend to an elevation of about 5,000 m in summer and descend to 3,000 m in winter.

8.2.7. Ovis Vignei (Shapo)

This species is found commonly in the valleys of Indus river. It measures about one metre and carries a massive type of horns which are generally extremely beautiful. It has sandy-grey colour and seldom goes beyond the snow-line. However, the meat is frequently taken by the villagers.

There are still a large number of wild animals found in Ladakh, for examples, Ibex, Marmot, and hare. Moreover, the region is having planty of birds (see in the text).

8.3. QUALITY OF LIFE

Quality is the essence of life. It may hold good for both animals and human-beings. It is attributed with a number of factors. Some of the factors are as follows:

8.3.1. Resources

The resources may be described as all the materials and processes in the world whether known or unknown. Materials may include minerals, soils, water, spaces and the processes are, gravity, radiation, evolution, biotic succession where through the energy is transferred. However, the resources may be concentrated and easily available and sometimes they are diluted. In the region, many wildlife times they are diluted. In the region, many wildlife resources have been eliminated. This creates the risk of environment.

8.3.2. Knowledge

This is a crucial factor which helps in determining the ability to use resources, in general and discover them in particular. This is the best way to estimate the quality of life and providing the material and spiritual forms of wealth which can be distributed among the people. However, knowledge is gained through, (a) experience, (b) scientific research, and (c) education.

8.3.3. Attitude

The manner of using the knowledge is much important than the availability of it. Public attitude is the best way to determine the knowledge. The attitude either stimulates or permits activities which either increases or decreases the quality of life.

8.3.4. Conservation

In the present context, conservation means to maintain the activities of renewable resources and prevent the waste of non-renewable resources, conservation is the best tool to determining the resources. This plays a vital role in determining the resources in the past and future too.

8.3.5. Population

All the four factors (resources, knowledge, attitude, and conservation) are used in determining the quantity and diversity of the material and spiritual wealth. In the same fashion, population also influence the quality of life. This may be due to the negative impacts on the conditions of resources and knowledge. The burden exerted by population tends to decrease the diversity of resources. Ehrlich and Ehrlich (1970) have given a good account of this factor in determining the quality of life.

Therefore, the quality of life, whether human-being or of the wildlife depends upon mainly five factors either in positive or negative sense. The equation of quality of life can be framed easily in taking into consideration the five determinants (modified after Bailey 1984).

$$QL = \frac{R + K \pm A + C}{P}$$

Where, QL = Quality of life.
R = Resources
K = Knowledge
A = Attitude
C = Conservation
P = Population.

Here, in the equation, attitude is a plus-or-minus factor for determining the quality of life, depending upon how it influences the efficiency of application of knowledge in achieving the aim.

8.4. WILDLIFE VALUES

Wildlife has great values in this materialistic world. They produce diverse values. There could be two types of wildlife values viz. (a) Aesthetic—which can not be measured in terms of money, and (b) Non-aesthetic—which can be measured easily in monetary terms. If, wildlife is put against the developments (that destroy the habitats), a thorough knowledge of wildlife values and methods of valuation must be known (Steinhoff 1971).

8.4.1. Kinds of Values

Several kinds of wildlife values have been recognised. King (1966) identified the following values:

8.4.1.1. ***Commercial*** : These values are the capitalized values of the income which are derived from the (a) selling, and (b) trading animals or their products. However, this depends upon the populations of wildlife. The commercial value may also include the value of wild meat and furs obtained from these animals. Therefore, the total commercial value of a wildlife population is the sum of capitalized values of whole business and the employee net incomes.

8.4.1.2. ***Recreational***: Wildlife is the source of pleasure, adventure, and increase physical and mental health. The people who derive the benefits from the wildlife may be hunters, fishermen, campers, hikers, photographers etc. The value, however, is measured by their respective willingness for outdoor recreation. The recreational value depends upon the local people's resources.

8.4.1.3. ***Aesthetic***: The aesthetic values are of historical significance. They form the part of poetry, art, and music. The people get the recreation from the songs of some birds. This is, however, dependent upon the identification of the wildlife.

8.4.1.4. ***Biological***: The biological values are contributed by the wild animals. They are part and parcel of complex ecosystems. The human-beings rely upon for food, fertilizer, water, aesthetic values etc. Therefore, the activities of wildlife provide enormous service to man, e.g. productivity and stability of ecosystems. There are a number of services which include pollination; soil tillage; natural regulation of plant and animal populations including cutting of diseased or inferior animals by predators; regulation of water resources etc.

8.4.1.5. ***Social***: There is an additional value of wildlife known as social. It is achieved when the other values go to the people. The people are benefited a lot through these values. The rural communities and less-developed organisations cannot be compared with cities and well-developed ones. Hence, social value is directly related to the quantum of other values.

8.4.1.6. ***Scientific/educational***: These are the values used for attaining the objectives of scientific study. The scientists (e.g. ecologists, physiologists, pathologists etc.) use the study of animals (wild) to extend their knowledge in the respective disciplines. However, the educational values are in the boundaries of schools, parks, in understanding the environment. The scientific studies have provided the bases for understanding human behaviour (Goodall 1965; Schaller 1972) and these values should not be under-rated.

8.4.1.7. ***Negative***: The negative values are the costs of wildlife damages to the crops and properties. Also, it accounts for the costs of controlling these damages. In general, the negative values are subtracted from the total commercial and recreational values of wildlife. There are many types of negative values, e.g. browsing by wild animals may kill the seedlings and tend to retard the forest growth and thus affect the forest industry significantly. Roosting birds may be nuisance due to their noise. Therefore, negative values are associated with all types of values of wildlife.

8.5. WILDLIFE CONSERVATION

Wildlife conservation is of utmost importance. There are several considerations:

8.5.1. Education

The conservation education consists primarily of helping people develop their appreciation of natural-resource values, from acquisition towards participation (Leopold 1949). The local people should be taught to use and enjoy their wildland resources. Wildlife management can be made more effective through education which aims at consisting of making land produce valuable crops of wildlife.

8.5.2. Problems

There are four major problems encountered with the wildlife valuation and the future of its conservation. Following are the main problems (Bailey 1984):

(a) A need to consider all values in decisions affecting the land-use. The wildlife values are not limited, rather, they are numerous and dispersed among many people.

(b) A need to develop sources of financial support for managing non-game and non-commercial populations of wildlife.

(c) A need for methods to compensate landowners for wildlife values produced and realized on private land, and

(d) A need for broader public awareness of the importance of habitat as necessary for sustaining the wildlife populations.

8.6 EXISTING SOURCES OF WILDLIFE

Since the inception of wildlife Advisory Board, Constituted under the Wildlife (Protection) Act of 1978, which has been modelled on the Indian Preservation Act of 1972, the Department of Wildlife Protection has revitalised sanctuaries and Reserves. Following are the existing sources of wildlife.

8.6.1. National Parks

Under the auspicious of wildlife (Protection) Act of 1978, a national park has been established in Ladakh region.

8.6.1.1. ***Hemis High Altitude National Park*:** This national park is situated in Leh and includes the catchments of the Markha and Rumbak valleys which independently drain into the Indus River at the village Nimmu and Phiang. In this national park, 8 species of mammals and more than 40 species of birds have been recorded.

Following are the salient features of this park:

Approach:

Nearest Airport	-	Leh (about 30 km from Nimmu).
By Road	-	Leh (about 35 km from Nimmu).
	-	Srinagar (about 300 km from Nimmu)
Nearest Railhead	-	Jammu (about 695 km from Nimmu)

Area of Park:

About 300 sq.km.

Accommodation:

Used for Camping sites.

Fauna:

The following fauna are met with in the area,

Ammon, Bharal, Read Fox, Shapu, Snow Leopard, Wolf, Hare, Marmot, and a variety of European and Upland birds.

8.6.2. **Reserves**

There are *two* types of reserves in Ladakh (cold desert),

8.6.2.1. ***Game reserves***: There are *four* game reserves (2 each in Leh and Kargil Districts).

(a) Shang (District Leh)

Area	:	80 sq.km.
Fauna	:	Bharal, Hare, Marmot, Red Fox etc. and there are more than 40 precious Chakores and Ram Chakares.

(b) Sabu (District Leh)

Area	:	15 sq. km.
Fauna	:	Bharal, Hare, Red Fox etc. and contains more than 40 individuals of Chakores and Ram Chakores.

(c) Kanji (District Kargil)

Area	:	100 sq. km.
Fauna	:	Hare, Ibex, Snow Leopard, Red Fox, Marmot, etc. and a number of species of pheasants and upland birds.

(d) Bodh Kharbu (District Kargil)

Area	:	15 sq. km.
Fauna	:	Red Fox, Hare, Ibex, and many species of bird which include the Chakores.

8.6.2.2. *Wetland reserves*: All the wetland reserves are present in the Leh District. The following are the Wetland Reserves:

(a) Chantan

Fauna	:	Bar-headed Goose, Pintail, Mallard, Gadwal, Coot, Shoveler etc.

(b) Summary

Fauna	:	Common teal, Coot, Bar-headed Goose, Mallard, Shoveler etc.

(c) Honlay

Fauna	:	Bar-headed Goose, Mallard, Pintail, Shoveler, gadwal, Coot, Common teal, etc.

(d) Chassul

Fauna	:	Bar-headed Goose, Mallard, Coot, Gadwal, Shoveler etc.

Noorichan

Fauna	:	Bar-head Goose, Mallard, Shoveler, Gadwal, common teal, coot etc.

8.7. MANAGEMENT

Wildlife management comprises of controlling the number, distribution, and quality of wild animals, either directly (as by manipulating hunting season) or indirectly (as by manipulating wildlife habitat). The basis of management of wildlife is, (a) selection of population, and (b) habitat parameter measurement. Selecting this data base is one of the most important management decisions. Therefore, the decision should be influenced by,

(a) the objectives of the management programme,

(b) the availability, cost, and precision of the methods for measuring population and habitat characteristics, and

(c) the budget of time and money for population and/or habitat analysis.

In Ladakh regions, all these factors are equally good. However, the data bases for wildlife management could be,

(1) Population indices;

(2) Population censuses;

(3) Measurement of habitat factors; and

(4) Indices of ecological density of the target population.

The population indices serve as a data base for many programmes of game management. These are also used to monitor changes in the abundance of non-game birds. Population census gives an idea of number of animals in a defined population. However, most wildlife populations are difficult and expensive to census. This is because of most animals are mobile and/or secretive. There are a number of census methods (Schemnitz 1980 and Caughley 1977). Following are the possible methods:

(a) Total Counts

(b) Sample Counts

(c) Population-ratio Methods.

The objectives of management may need information on the quantity, quality, distribution, persistence, dependability or the level of use of any food or cover resource in the habitat. The ecological density of a population is the number of animals relative to the quantity and quality of habitat resources available to the animals.

8.7.1. Future Planning

There is an urgent need to frame future planning for maintaining the wildlife. Wildlife is much associated with the development of forests. Following are some of the guidelines ((Sagwal 1991):

8.7.1.1. ***Forest diversity***: There are *three* basic requirements of habitat of wildlife, (i) Food, (ii) water, and (iii) cover. This is an observation that when an area provides a diversity of habitats, wildlife population remains stable. However, the forest diversity is much influenced and altered when disturbed by, (a) wildfires in the area, (b) Harvesting, and (c) pests. Therefore, forest diversity must be maintained for the prosperity of wildlife.

8.7.1.2. ***Tree species composition***: It has been observed that some wild animals are primarily associated with softwood stands, whereas, others are adapted to the hardwoods. However, mixed woood/forest stands are always preferred by several animals. In Canada, for maintaining a balance of wildlife, a ratio of 65 per cent /35 per cent softwood/hardwood is kept frequently. But there are some problems in other areas to maintain this ratio.

8.7.1.3. ***Forest succession***: There are several stages of succession and each provides the wildlife a different habitat condition. However, the successional stages can be grouped into,

(a) *Forest openings*: This comprises of the grasses and herbaceous plants.

(b) *Shrub-sapling*: This provides woody browse and may consists of fruiting shrubs and the trees.

(c) *Pole timber*: This is less valuable but provides food and cover to wild animals.

(d) *Mature timber* This provides winter cover by blocking winds and reducing snow-depth.

(e) *Old growth*: This is used for nesting sites for birds and the wild animals for shelter.

8.7.1.4. ***Water courses***: Provision of water courses in the wildlife habitat is the prime factor for their survival. The lakeshore or streambank is the best way to provide drinking water to animals. Moreover, trees and other rooted vegetation along a water course stabilize the soil, preventing erosion etc.

8.7.1.5. ***Corridors/Deadwoods***: The uncut strips of forest stand crossing a clearcut edge and provide 'wildlife corridors' from one habitat to another. However, the hard corridor provides excellent firebreaks. The deadwoods provide a good source of habitat. The following kinds of trees are kept for use:

(a) *Snag*. It is a standing dead/dying tree. It provides feeding grounds for some insect-eating birds and the nests also.

(b) *Cavity trees*. A cavity is one that attains a diameter (at breast height) more than 30 cm. They provide holes and the cavities which are useful for some birds.

8.7.1.6. ***Policies***: It has been noticed that some of the wildlife species are going to extinct and there is a great danger to the ecosystem. Therefore, there is an urgent need to save and preserve the wildlife. A strong policy must be enforced for the safety of wildlife in the cold desert.

It is concluded that Ladakh (cold desert) though has the forest worth the name, the wildlife is found in abundance. The wildlife is of much importance in the state's revenue. Wildlife is very essential in maintaining the unstable ecosystem. The wild animals have some common features. There is much need to maintain the quality of life and many factors contribute this. They also provide 'values' which are the essence of life. The region has several National park, Reserves which are the homes for wildlife. The management is of utmost importance for conserving and preserving this great asset. Therefore, every sincere effort must be done for the cause of the development of wildlife in cold desert.

Chapter 9

Medicinal and Aromatic Plants

Ladakh region is not only bound with essential oil plant wealth but has a great potential for future also. A large number of medicinal and aromatic plant species are found and hence a thorough survey of the area is urgently needed to explore and assess the valuable forest wealth. The area experiences the difficulties of extraction and labour shortage. However, to start with, the cultivation of medicinal and aromatic plants can also be taken up in the forest plantations. The potential of renewable wealth (medicinal and aromatic plants) has not been judiciously exploited so far. Accordingly, an attempt is urgent required to explore the possibilities for cultivation of different types of essential oil plants in the region.

9.1 EXTENT OF ESSENTIAL OIL PLANTS

It is well known that forest products other than wood are prescribed as Minor Forest Products. Among these, essential oil plants are very prominent. This little variety of plant wealth, however, has a big potential in the region. Therefore, several useful and essential oil plants have been identified which constitute this flora of Ladakh (Table 9.1). The beauty of the region is that whatsoever little vegetation it has, the proportion of essential oil plants is much more but so far it has been the hidden wealth. However, the region is divided into several sub-regions (Singh 1970):

9.1.1 Changthang

This region is quite vast, broad and experiences more and strong winds. This is blessed with a number of lakes, rivers (namely Sindh), and possesses a rich deposits of borax and sulphur.

9.1.2. Drass-Suru

This represents the catchment area of Drass and Suru rivers. This region has been recognised as the richest belt due to its luxuriant alpine vegetation.

9.1.3. Leh-Khalsi

This is a very prominent region. It lies in the catchment of the main Indus river.

9.1.4. Nobra

There are two famous rivers viz. Shyok and Nobra rivers. The region comprises of variety of vegetation.

9.1.5. Wakha-Mulbik

The soil of the area is richer due to the moist conditions found in very high frequency.

9.1.6. Zanskar

The Zanskar nalla (nullah) runs through the region which is the main drainage of the tract.

9.2. HOW TO EXPLORE

It is devoid of forest vegetation but rich in medicinal and aromatic plants. It is a hidden resource and has a great potential which can be exploited commercially and economically. A few of the ways are as follows:

9.2.1. Research Needs

The area comprises of a rich material of essential oil plants. But it has not been exploited so far due to one or the other reasons. The exploration of this rich possession can now be taken up, if, the research work is carried out in the right direction and therefore, priority areas of researches could be:

(a) A detailed survey of the whole region with regard to essential oil plants.

(b) Preservation and maintenance of the germ-plasm.

(c) Agrotechnology of important essential oil plants should be evolved to popularise their cultivation on commercial scale.

(d) Economics of collection, cultivation, preservation, exploitation must also be calculated.

9.2.2. Quantity and Quality of Plants

Due to rich essential oil plants, the need of the day is exploitation. All the plants of aromatic nature are of good quality which are very famous for their endless uses. Hence, both the factors (quantity and quality) are favourable and the market can depend upon them upto a great extent with safe.

9.2.3. Demand of Market

The market plays a vital role in setting up of some industries and/or to take up some enterpreneur on commercial scale. However, a small supply of produce can be consumed within the country. But the demand of the market is favourable as our country is running short of these important drugs. There should be ample scope of creating more facilities for marketings for such products.

9.2.4. Availability of Labour

In the region, labour supply is a major problem. The wages are very high due to inaccessible area. For a successful programme, labour supply can be meted out from outside. This type of practice is being followed by the BEACON Project. So, the labour supply should not be a hinderance in the exploitation of such plants.

9.2.5. Transport/Communication Facilities

All the sub-regions of the area are well connected except a few. However, the road communications between the region of Ladakh and Kashmir Valley remain in operation for only 5-6 months a year. It should not be considered as limiting factor in the transportation of aromatic plant wealth. The help of trucks should be taken up.

9.2.6. Management Practices

The possibilities of starting small types of factories for manufacture of crude produce be explored. This will certainly curtail the cost of transportation of raw material outside the region. The collection of drugs must be based on cooperative basis. The medicinal and aromatic plant areas/fields require enough protection against biotic factors. However, the browsing of goats should be banned. This will give more chances for right exploration of this valuable wealth.

9.3 FORESTRY EXTENSION

The only alternative to explore the idea of expanding the scope of medicinal and aromatic plants is to boost the intensive forestry. It can be taken up on sounder footing with the help of implementation of forestry extension. Forestry extension is the continuous process designed to make rural people aware of their problems relating to forests and indicating to them the ways and means by which they can solve them (Sagwal 1981). It is not only the education of the rural people in determining their problems and methods of solving them in relation to the forest but also inspiring them towards positive action in doing so.

9.3.1. Aims of Forestry Extension

An effective programme of educating the people of Ladakh with regard to the following is essential to tackle the whole gamut of the medicinal and aromatic plants problems (Sagwal and Gupta 1988).

9.3.1.1. ***Cultivation***: In hilly areas, the people make extensive use of medicinal and aromatic plants whereas they lack knowledge and initiative for the proper survey and growing of these plants. Hence, to save this wealth from destruction, the farmers need to be educated. The Extension Educationists have the great and challenging task before them to pursuade and motivate the farmers/growers to cultivate the economic plants with the improved methodology and technology.

9.3.1.2. ***Restrictions to over-grazing***: The people keep a large herd of cattle, most of which are useless and results in unnecessary pressure and burden on grazing lands. This causes the problem of erosion and loss of valuable plants. This is subject to the strong approaches of Forestry Extension. Reddy (1983) reported that free grazing is allowed and hence more number of cattle are grazing than the carrying capacity of the land. This is a direct invitation to several types of problems, e.g. erosion.

9.3.1.3. ***Fire control***: Generally in summer season, the areas with vegetative cover are susceptible to the fire hazards. The fire destoys the seedlings and saplings (including the medicinal and aromatic plants). The left over-litter which otherwise would have decomposed on the floor, gets burnt and washed away due to wind or the rains (in some cases). Fire control measures should be taught to the villagers which is possible under Forestry Extension Education Programme.

9.3.1.4. ***Imparting skill***: The training of the concerned officials and the rural people is very essential. They should be well conversant with the latest know-how of the medicinal and aromatic plants. However, the skill should be gained through the Forestry Extension.

9.3.1.5. ***Changing attitude***: It is very essential to change the attitude of the farming community. Now, the time has come to utilize each and every centimetre of our land in any form (e.g. agriculture, forestry, cultivation of medicinal and aromatic plants). There should be a change in the attitude and people should be pursuaded to shift from 'Mono' to "Tripolar System" of Farming to meet the diversified needs of the people of the area (Gupta 1983).

9.3.2. The approaches

The education of the farmers will pave the way to boost this wealth of the region. Therefore, in achieving the goal, the involvement of the youths is of paramount importance (Sethi 1977). The extension approaches as far as possible should be (a) sound, (b) acceptable to the rural, and (c) easy to follow.

Every effort should be made to create the consciousness among the people. Following methods of extension should be taken into consideration.

9.3.2.1. ***Seeing is believing***: To win the confidence of the people, 'Demonstration Plots' must be established, wherever possible. The Forest Department should undertake and encourage the entrepreneurs for commercial plantations of the chosen medicinal and aromatic plants which is easy to approach even for a small farmer.

9.3.2.2. ***T & V system***: The T & V System be employed for educating the people of the area. The new approach requires working with the farming community selected from the socio-economic groups and aims at reaching to large number of farming families quickly (Singh 1981).

9.3.2.3. ***Intensive publicity efforts***: This comprises of the preparation and distribution of, (a) Brochures, (b) Pamphlets, (c) Posters etc. This gives the exposure of the rural people to the audio-visual publicity through Cinema, Documentary films, Newspapers, Radio, and the T.V. talks. This will expect to create a favourable tempo among the people (Trigotra 1983).

Hence, medicinal and aromatic plant wealth is a good source of revenue but has not been exploited judiciously for the betterment of the rural poor. Still, there are some possibilities to explore this wealth in the right direction. The only need is to create the consiousness among the people of the area. Forestry Extension plays a vital role in exploiting and increasing the extent of the medicinal and aromatic plants. In achieving the aim of forestry extension, very serious and sincere efforts are needed.

Table 9.1: Essential oil plant wealth of Ladakh region

Botanical Name	*Family*	*English/Hindi*	*Description*	*uses*
(1)	*(2)*	*(3)*	*(4)*	*(5)*
Artemi Sia Brevifolia Linn.	Compositae	Wormwood (E) Valaitiof santin (H)	Perennial under shrub, 1/2 to 1m high, leaves 2.5 cm long, whitish in colour.	Contains essential oil (0.7%) having santonin and also artemisin. Used as vermifuge and also use ful as cardiac stimulant.
A. Maritima Linn.	Compositae	Wormseed/Santonica (E) Kirmala (H)	A stout, much branched, perennial aromatic shrub collected when flower heads not fully matured.	Drug used for expelling worms from the stomach, also used in fevers, dropsy and as a stimulant.
Carum carvi Linn.	Umbelliferae	Caraway (E) Shiajira (H)	A herb found wild in the whole region	Contains essential oil (45-65%), used as condiment, and carminative oil from fruits is used for mouth-wash and perfumes.

(1)	(2)	(3)	(4)	(5)
Chenopodium Album	Chenopodiaceae	Pigweed/lambs quarters (E) Bathus (H)	A common and strongly Scented herb.	The plant contains essential oil, carotene vitamin C, used as an anthelmintic, leaves and tender twigs used as vegetable.
C. Ambrosiodides Linn.	Chenopodiaceae	American worm seed (E)	A common and strongly scented herb.	Yield of oil from dried fruit comes out to be 0.75-1.60 per cent. Used as an anthelmintic, leaves and tender, twigs used as vegetable and fodder.
Datura Stramonium Linn.	Solanaceae	Thorn apple/Jomsonweed (E) Dhatura (H)	Bushy annual attaining a height of 1.5 m. Stem and branches are found, smooth and green. Flowers are pretty, large and white. Fruit is about 3-4 m long.	Essential oil (0.45%) extracted from the leaves the Chief alkaloids present in this drug are hyoscymine and little atropine.

(1)	(2)	(3)	(4)	(5)
Dioscorea Deltoidea Wall ex. Kunth	Dioscoreaceae	Potato Yam (E) Gaithi, Ratalu (H)	Stem climbing leaves cordate, acuminate, flowers 2 mm in dia solistary or in clusters.	Contains essential oil (Diosgenin 4.8%), used in prepartion of cartisone and other steroid hormones, effective in treatment of rehumatoid, arthritis.
Mentha Arvensis Linn.	Labiatae	Field mint (E) Pudina (H)	The plant is 15 to 30 cm high with hairy glabrate.	The essential oil varies from place to place. The leaves with sweet smelling nature are source of essential oil mentol or pepperment oil. The piperita spp. is famous for its use for headache, rheumatic pains, cough drops, mouth washes etc.
Podophyllum Hexandrum Royle	Podophyllaceae	Indian Podophyllum (E) Papra (H)	Succulent erect herb with creeping stock, leaves 15 to 25 cm in dia. divided	Produces essnetial oil 11.1% resin). Contains a resin podophyl slum, used in

(1)	(2)	(3)	(4)	(5)
			into 3 or 5 lobes, flowers white or pink.	many skin diseases and tumerous growth. Also used in cancerous growth.
Saussurea Lappa C.B. Clarke	Compositae	Costus (E) Kuth (H)	A perennial herb, 1.5 to 2 m high, leaves, flowers bluish-purple or almost black.	Contains 1.22 per cent essential oil. The essential oil contains 1 per cent terpentine, 20 per cent aplo taxene, 60 per cent secquit-erpines. Saussrurine found in the root. The roots are used in woollens for its insect repelling, has antiseptic and disinfectant properties.

Chapter 10

Combating Desertification

The Himalaya is one of the most endangered ecosystems on the earth and every effort is required to preserve this vast resource, where more than one-third of the total population of the subcontinent is dependent. Ladakh-cold desert, forms an integral part of the great Himalaya. It is the loftiest inhabited land of the World. Hassnain *et al.* (1977) stated that Ladakh is for the most part a desert of bare crags and granite dust with vast arid table lands of elevation. The problem of desertification is on and may lead to further disasters. This can pose a severe threat to the ecosystem. Therefore, the present attempt is in this direction to explore the possible methods, ways and means and the appropriate techniques to fight against this advancing nuisance problem.

10.1. WHAT IS DESERTIFICATION

Desertification has been explained in different meanings bringing out almost the same theme. Many workers have defined desertification (Hellden 1988). A few of the important definitions are as under:

(a) Desertification is the diminution or destruction of the biological potential of the land, and can lead ultimately to desert like conditions (Anon.1977).

(b) Desertification is a process of improvishment of arid, semi-arid and sub-humid ecosystem by the combined impacts of man's activities and drought (Dregne 1976).

(c) It is defined as a process leading to reduced biological productivity, with consequent reduction in plant biomass, in the lands carrying capacity for livestock, in crop yield and human well being leading to the intensification or extension of desert conditions (UNCOD A/CONF 1977).

(d) Desertification (also called Desertization) is defined as the spread of desert-like conditions in arid or semi-arid areas due to man's influence or to climatic change (Rapp 1974).

The cold desert (Ladakh) is a vast area and is a victim of desertification. To maintain the ecological balance, control of desertification is utmost important. Before taking any initiative in combating this nuisance problem, following basic principles should take into consideration.

10.2. BASIC PRINCIPLES

The process of desertification has taken up a grave situation not only in the cold desert of Ladakh but the World over. It is estimated that desertification is slowly but steady turning productive land to merely desertic conditions at a rate of some 60,000 sq.km.yr^{1} (World basis). Combating of desertification is guided by the following basic principles (Anon. 1988):

(a) An integrated system approach is required to bring together technological, economic, socio-cultural and ecological aspects.

(b) Stress must be placed on a field-level, action-oriented, bottom up approach for transferring information to locals.

(c) A genuine attempt should be made to synthesize local knowledge and experience with scientific resources from other community.

(d) The approach should be developed for the solutions that offer improvement in life quality.

(e) Emphasis should be laid on collaborative approach between the Government/Agencies and the local people.

10.3. FIGHT AGAINST DESERTIFICATION

For preventing/controlling desertification, coordinated complex of social, economic and technical measures is required. Following are some of the methods/ways to fight against desertification for controlling it effectively.

10.3.1. Technical Measures

A high technical knowledge is needed to take up steps. It requires expertization. A few of the considerations are as under:

10.3.1.1 *Soil/water conservation*: It is performed by two ways:

10.3.1.1.1 *Prevention of soil erosion*: The soil erosion is the function of, (a) erosivity of rainfall, and (b) erodibility of soil and land features particularly the slopes. The scanty rainfall causes more wind erosion. This leads to the robbing of the richest top soil and the crops are burried in drifting debris. This is a great hovoc in the region which leads to the unstability of ecosystem. It is a fact that loss of soil is high in the bare fallow (naked surface) and least under vegetation (grass) cover. Mann (1982) observed the cumulative sand removal from the bare sandy soils was 9.66 cm whereas no removal occurred on the soil surface covered with grass. The following measures could prove to be effective in checking soil erosion:

(i) *Contour farming*: on a 4 per cent slope, the contour farming reduces the run-off drastically in comparison to up and down farming (Tejwani 1974). Also, contour bunding is effective in desert areas where the soils have high infiltration rates. These practices should be popularised in cold desert (Ladakh) which will prove to be boon to dwellers.

(ii) *Wind breaks/Shelter belts*: Wind breaks and shelterbelts play a vital role in desertification. They protect the land from wind erosion in sandy soils. This acts in the absence of crop residues and mulches. Pathak (1958) has established the fact that wind breaks/shelterbelts in the form of agroforestry (old name farm forestry) could be helpful in controlling the run of soil erosion.

It has been noticed that the trees of *Salix* spp. and *Populus* spp. give bushy growth but do better near the water courses. They serve in many ways viz. protect the farm land, provide fodder, gives good source of fuelwood and can be used for several other functions.

(iii) *Terracing*: It is much useful in sloppy lands. These are generally used for controlling erosion and conserving rainfall. The ploughing should be done parallel to the terraces. However, the terraces must be supplemented with some good cropping practices. The grasses like clover, alfalfa can provide a good protective cover to the terraces. If, possible on the slopes, retaining walls must be erected to stabilise the terraces. In cold desert, this can protect the area to a great extent and will help in stopping desertification.

10.3.1.1.2. *Water Harvesting*: Water harvesting pertains to the process of collecting and storing water from the specified areas. The water can be stored successfully in tanks, reservoirs and in the soils. The water is diverted from the watershed (small in nature) to the fields/crops. This practice, however, was popular in many countries of the world particularly in the desert regions. These old systems of water storing are the pavements of modern water harvesting systems.

The condition of cold desert is so deplorable that even the driest crop like barley can not be grown without irrigation. The best way of harvesting water is to construct the ponds on the hills and divert water of glaciers. The size of the storing tanks may be according to convenience and the flow of water. This water can be utilized judiciously for minimal life saving for increasing crop production. Tejwani and Bhardwaj (1982) reported an increase in field crops of wheat (2,143 to 4,119 kg ha^{-1}) and barley (3,644 to 4,725 kg ha^{-1}) with 2 life saving irrigations of 4 cm. Moreover, Ladakh gets the dry snow which is a problem in itself as it drifts away with the winds. This is because of the powdery nature of the fallen snow. It is observed that this nuisance can be controlled if, fields are ploughed immediately after the snowfall. This will facilitate the work of conserving moisture in the soil which can be retained for a longer period. Therefore, water harvesting should be popularised on large scale.

10.3.1.2. ***Development of pastures***: The pastures play an important role in stabilizing the vulnerable conditions of the region. There are two main requirements for the development of pastures in the cold desert viz. (a) A sufficient supply of pasture plant minerals. This facilitates the enough growth of the plants (may be grasses) and can provide forage for livestock, and (b) During grazing season, the following should be regulated to ensure better performance: (i) number and kinds of livestock, and (ii) periods of grazing.

Fom Zoji La to Drass area of Ladakh, the pastures can be developed due to great potential. The following are the few grass species which are doing well in Ladakh (Table 10.1).

Table 10.1: Grasses suitable for the region

Place of suitability Tested	*Grass species*
Drass	Tall fescue, alfalfa (yarkand type), red clover, lucerne, lenbil, oats, alfalfa (local)
Leh	Clover, lucerne.

The following are the observations in cold desert regarding the pasture development.

10.3.1.2.1. *Overgrazing*: The livestock needs sufficient lands for their sustenance, mainly the grazing lands (pastures). Overgrazing exhibits several harmful effects on the soil/ground (Reddy 1983) viz.

Firstly, due to overgrazing, the ground floor gets hardened. This results in less percolation of water and more run-off causing the soil erosion.

Secondly, damage to natural regeneration as the seedlings are damaged/destroyed.

Thirdly, grazing reduces the grass cover leaving the ground bare, and

Fourthly, non-availability of grass for wildlife and sometimes graziers burns the grass cover to get good grass in the ensuing season.

10.3.1.2.2. *Rotational Grazing*: There is a dire need to regulate grazing. Therefore, rotational grazing must be followed. This will help in restoring the depleted pastures, otherwise erosion and soil losses are heavy and proves to be a severe threat to the ecological balance. Grazing should be regulated through the following measures (Anon. 1972):

I. *Indirect measures*

(a) Controlling livestock population,

(b) Imposition of suitable fee, and

(c) Alternative to pasturage.

II. *Direct measures*

(a) Deferred grazing,

(b) Closure of grazing,

(c) Limiting grazing incidence to grazing capacity, and

(d) Penning cattle in interior forests.

10.3.1.3. *Strong landuse planning*: A strong landuse planning is very essential for the development of desert. There are several limitations in the way of this planning e.g. coarse textured soils, erratic climate, and a short span of growing period. The fodders and vegetables can conserve and improve the soil conditions. The fodder grasses like clover, alfalfa must be grown on large scale. They will be fed to the animals and act as protective cover. For boosting the economy, a proper landuse planning is needed on priority basis. To a great extent, the vegetables like cauliflower, cubbage, peas can be grown as the market potential is available within the region. However, there are some problems in the way of strong landuse planning and a few of them are as under:

(a) Unreliable Weather,

(b) Small size of land holdings,

(c) Lack of proper landuse policy,

(d) Paucity of resources,

(e) Lack of bottom up approach etc.

The Government and the Development agencies should chalk out the future plan of strong landuse planning. This will give a judicious use of land.

10.3.1.4. ***Irrigation facilities***: Irrigation is the prime factor for raising crops. It has more importance in the cold desert because of paucity of irrigation water. The land productivity can be improved to a great extent, if, adequate irrigation facilities are created and maintained in a proper balance. It is expected that with the commissioning of Stakna Project and Igophy Canal, about 4,000 ha land could be given irrigation. The cold desert is blessed with mother nature's precious gift of River Indus which is supposed to be ninth longest river with a discharge capacity of about 5,000 gallons/sec. If, water of this river is utilized properly, it can convert the yellow tract into greenery.

Now, with the advancement of technology, new methods of artificial irrigation are available which can be tested under cold desert conditions. Sprinkling is one of the best methods of new technology which can prove to be a boon to the area. This method gives uniform wetting of soils, protect the soils from their erodibility and will increase water use efficiency which is more economical in comprise to conventional methods of irrigation.

10.3.1.5. ***Increasing soil fertility***: This is attributed with the farming practices and management of water and its deficiency is the salient feature of the region. The fertility of soil can be increased by adopting the following measures:

(a) Mulch farming,

(b) Zero tillage,

(c) Growing of legumes,

(d) Use of plastic in agriculture,

(e) Application of fertilizers,

(f) Re-seeding of pastures,

(g) Reducing run-off of water.

10.3.2. Integrated Approach

The integrated approach is the demand of the day. Now, the time has come not to go by single approach rather integrated all possible approaches so that maximum facets of the problems of desertification can be tackled at the shortest possible period of time. This approach includes the expertisation of the scientists/workers of the following disciplines:

- Environmental Sciences,
- Ecology,
- Soil Science,
- Plant/animal breeding,
- Geology,
- Agronomy,
- Meteorology,
- Forestry/agroforestry etc.

A few example are as under:

(a) Irrigation experts (agronomy) may exploit the water of Indus river judiciously for hydro-electric projects. The services of engineers are also essential to make such projects a success. He (the irrigation expert) can advise the farmers/resource-poor personnels for proper and justified use of irrigation water to the soils.

(b) The meteorologist can predict the weather forecasting. For this purpose, long term meteorological data are needed.

(c) The expert of soil science can locate the suitability of land for crop production or other use. Similarly, based on the experience and expertisation in soil science, the best possible use of land capability classification would have been done consciously.

Similarly, the functions of other disciplines can be explored and exploited in the true sense. If, this approach (integration) is taken up seriously and with true spirit, then there may be a drastic change and development in the cold desert.

Forestry extension can fulfil the cherished aims of combating the desertification. For a social scientist, extension is a behaviouristic approach to human problem which aims at changing their 'attitude and practice' (Mukhopadhyay 1971). Although forestry extension seems to be an easy approach, it is not as one sees it. It is a very combersome task. Khan (1987) suggested various means of carrying out extension work. A few of them are as follows:

(a) individual level contact,

(b) group or community level contact,

(c) proper use of mass media, and

(d) demonstration area/plot.

The integrated approach can be implemented with a strong extension technique. The recent system of 'Training and Visit' can show the way for tackling the problem on large scale. This system has the following functions:

(i) Field visits,

(ii) Training of field staff, and

(iii) Related training and exposure to research.

There are a number of agricultural extension models. These can be modified for controlling the desertification in cold desert. Hence a moderate type of forestry extension model should be developed so that the same can be utilized for the noble cause of combating desertification.

10.3.3. Administrative Measures

Desertification in cold desert is a grave situation caused by several agencies. In addition to the technical measures and integrated approach to tackle the problem, administrative measures are the pivot in this programme. The following administrative measures can be taken up in the present context. It will pave the way to activate the people in controlling the biggest nuisance of the region:

10.3.3.1. ***Creating awareness***: Through this, the locals can be stimulated and mobilised in taking up the task of combating the problem of desertification. There are two most effective tools in creating the awareness among the people viz.

(a) *Extension Agencies*

The following are some of the functions of extension agencies:

(i) consciousness among the local people about the harmful effect of desertification;

(ii) tell the uses of trees/shrubs/grasses in controlling the nuisance problem;

(iii) train the village level workers, voluntary agencies or some other individuals involved in motivating the people to be aware of desertification, afforestation programmes etc.

(iv) help the local communities in organising cooperatives or self-help groups and contribution of labour in establishment and management of the works regarding controlling desertification.

(b) *Communication media*

An intensive publicity campaign is required when extension services do not reach to individual. Therefore, communication media is the best way to spread the message. The most popular communication media are radio and television. Radio can reach even to the remotest areas. This proves to be the most effective media. In addition to these two vital tools, there could also be other media such as,

(i) film shows;

(ii) preparation of leaflots;

(iii) pamphlets;

(iv) brochures;

(v) published material;

(vi) posters;

(vii) newspapers and magazines.

10.3.3.2. ***Role of institution***: The relationship between various dispciplines (like agriculture, forestry, animal husbandry etc.) and characters of social institutions and customs is so intimate that it is very difficult to decide which is the cause and which is the effect (Buchanan 1959). The institution has a significant role to play in controlling the process of desertification. The State Government has established an 'Antidesertification Agency' which performs the following functions:

(i) Planning;

(ii) Programming; and

(iii) Monitoring.

The utmost important function is to identify, formulate and implement the projects in the area. However, these depend upon the principles of sound ecology. Therefore, the answer to desertification problems does not necessarily lie in the technical or integrated approach but in the people-oriented development along with the establishment of helping institutions.

In conclusion we can say that Ladakh though blessed with many gifts of nature, faces the acute problems of desertification. Combating desertification, five principles have been taken as the base. There is an urgent need to explore the possibility of the methods of controlling the desertification. If, all the measures are implemented with full vigour and in true spirit, then there is no reason why we cannot tackle the problem. Several possible measures for combating the desertification have been discussed in detail which will show the path in right direction. The role of administrative measures should not be underestimated.

Chapter 11

Afforestation Challenges

"Plant and protect trees for a better environment and prosperous future"

—*A Slogan*

Ladakh (Jammu and Kashmir) is like a crown for the nation as geographically, it is studded in the great Himalaya. The forests in the hilly states play a significant role in the life of the people. Because the forests, in real sense, serve as the very source of existence for the hilly folk and help in sustaining their difficult and hard life. But unfortunately, this region is devoid of forest cover which is otherwise very essential. However, the people have become conscious about the forest wealth and a deep sense of awareness is dawning in the rural areas about the urgent need for promotion and protection of forest wealth.

The afforestation works have too been limited to these areas. This resulted into the fringe forests. Therefore, the impact of these fringe forests, on the improvement of environment remained confined the valleys and depressions. It is obviously true that these are not sufficient to meet the growing demands of fuel, fodder and small timber etc. Seeing all these cases, there seems to be good scope of afforestation in the region and needs to be extended over the whole of the area. In this region, afforestation, if, not impossible, but seems certainly very difficult task. Accordingly, the present attempt is in this direction, to explore the possibility of undertaking the afforestation programmes.

11.1. AFFORESTATION ACTIVITIES

There are several ways through which afforestation works are carried out in the region. Therefore, it is considered in the following manners:

11.1.1. General Activities

Now, every possible effort is being made to take up the plantation work on large scale. This leaves a very short growing period for the plants in such conditions, places a great setback to the afforestation programmes. The most common species of forest trees, generally taken up by the farmers, are *Populus* spp. and *Salix* spp. Both the species are easy to propagate through cuttings. As such no research work has been conducted by Forest Research Institute, Dehra Dun (Mathur and Sharma 1983). A plan scheme on "Cold desert afforestation", however, was included in the plan for field research. It is apparent that whatever knowledge is available, is the result of experiences gained in the afforestations carried out by the J & K State Government.

11.1.1.1. ***Objectives***: Afforestation brings the new plantations in the area. The immediate objective of this programme is to cater the local needs/ demands for fuelwood and small timber for constructional purposes. The wood in any form is scarce in the area and, therefore, there is an urgent need to investigate the adaptability of some exotics so that the local demands may be fulfilled within a short period of time.

11.1.1.2. ***Some possibilities***: The region is lacking the facilities of irrigation either through natural or artificial sources. Hence, for afforestation, minimum irrigation facilities should be created. It is supposed that following exotic species may give good response in growth than indigenous species (Singh 1977).

Populus euphratica,

P. alba, and,

P. candicana are the species of Poplars which have been already tried in cold desert area of Ladakh. Moreover, there is great possibility of the following species of Poplars, which are supposed to do better,

P. laurifolia (South Siberian),

P. lasiocaspa (Central/Western China),

P. suaveolens (Central Asia),

P. tachamahaca (Alaska, Canada),

P. tremuloides (Alaska to new Foundland)

All these species of poplar are very promising under cold conditions and should be tried under the cold desert conditions.

In addition to these poplar species, following *Eucalyptus* spp. could also be tried (Table 11.1).

Table 11.1: *Eucalyptus* spp. suitable for varying temperature

Temperature tolerance	*Species of Eucalyptus*	
-18°c	*Eucalyptus*	*parviflora*
	E.	*niphophila*
	E.	*gunni*
-15°C	*E.*	*Coccifora*
	E.	*johnstonii*
-12°C	*E.*	*aggregata*
	E.	*dalrymphleana*
	E.	*delegatensis*

11.1.2. Activities of FRL of Defence Ministry

The Field Research Laboratory (FRL) of the Defence Ministry in Ladakh (at Leh) has brought about a sea change in the afforestation campaign, development of agriculture and animal husbandry (Anon. 1991).

The FRL is situated at an altitude of about 3,740 m (a.m.s.l.) and working since 1962 for developing agro-animal resources. But later on diversified to improve the ecological balance, mainly through afforestation programmes. The laboratory has planted the following tree species:

Forest Tree species : Poplars and Willows.
Horticultural species : Apricots and Apples.

Besides, the FRL has taken up the production of vegetables. Similarly, in the field of animal husbandry, the FRL has taken steps to improve the quality and quantity of milk, meat, and eggs.

Now, in addition to these activities, people have started planting trees on the boundaries of their fields. Several woodlots have been erected near the villages. Different species of trees have been tested with different altitudes (Table 11.2).

Table 11.2: Tree species adapted to different elevations

Elevational height (m)	*Forest tree species*
upto 3,400	willow, poplar, robinia,
3,400 to 3,900	willows and Poplars
3,900 to 4,100	*Salix fragilis*

11.1.1.3. ***Growth Pattern***: The pattern of growth of plants also vary from place to place. It has been observed that with the increase in height (altitude), there is a gradual decrease in the growth of plants. Willow and Poplar plants which have shown the growth rate of 4.25 m at the height (altitude) of 3,300 m during a period of five years are only 2.50 m tall at the height (altitude) of 3,800 m (Negi 1985). And, similarly, the growth of plants are better within the habitations.

11.2. Changing into People's Movement

The major land area in desert is rainfed agriculture land. Therefore, there seems to be-maximum scope to carryout a massive tree planting programme on such lands. The following steps must be taken:

11.2.1. Planting along the Boundary

The trees of poplars and willows are planted along field boundary in lines 3 m apart at a close spacing of 1m. This type of planting serves as a good fencing. It will help in increasing the yield of agriculture crops. Also, farmers get sufficient fuelwood.

11.2.2. Planting Rooted Plants

It is a normal practice in the region that the branch cuttings, as such, are planted. The length and thickness (diameter) of the cuttings may differ from person to person as no standard measures are taken. This is, however, a wrong practice which takes too much time in sprouting the stumps (cuttings) and moreover, may lead to much failure due to the unadaptability of plants in different type of sites. Keeping this in view, now the people are encouraged to plant the rooted-cuttings of the trees. This gives more chances of survival. However, the Poplars and Willows are easy to root but the cuttings take too much time in attaining a good size.

11.2.3. Fodder Trees

It has been noticed that there is a scarcity of green fodder to feed their animals. Therefore, modern scientific technology (by installing a solar photo-voltaic pump) should be adopted for fodder multiplication. There are some species of Poplars and Willows which can be used for fodder purposes and similarly, the introduction of Robinia (a legume) can pave the way for reducing scarcity of green fodder. In addition to these, some species of grasses like orchard grass, bromus, clover and lucerne are the promising species of fodder which can successfully be tried. If, these efforts are taken with full strength and vigour, the livestock can be reared easily which can meet the ever-increasing demands of milk etc.

11.2.4. Kisan Nurseries

The Forest Department of J & K State and the development agencies should lay emphasis on starting Kisan Nurseries. This will help the farmers in several ways, viz. (a) gives employment opportunities, (b) will keep them busy. The Government should make provision to start Kisan Nurseries in each and every village. The material (seeds and cuttings of trees etc.) should be provided to the farmers by the Government agencies. It will lead to make afforestation programme a people's movement. In the starting the setting up of nurseries may be on small scale and with the pace of time and the interest of the people, it must be taken up on large scale which can cater the growing needs.

11.2.5. Forest Forum/Club

Though people have great love for trees, there is still an urgent need to make them more conscious about the forests/trees and their importance in human life. For achieving this, some forest forums/clubs must be established where the local people can be given good ideas about the forests. This will certainly lead to adopt this profession in true sense. Therefore, every sincere effort should be made to achieve the set norms of afforestation.

11.3. EFFECTING ECONOMY IN AFFORESTATION WORK

Economy is the crucial factor in any developmental activity. If, sincere efforts are done, then it can be easily achieved. Therefore, effecting economy in afforestation works can be viewed as under:

11.3.1. Elimination of Fencing

The work of afforestation in cold desert is a costly and difficult process as it involves a lot of labour/money in resorting the barbed wire fencing. Though the practice is old but can be seen oftenly. Now, some alternatives/substitutes can easily be provided. Live fencing can be the ideal one, in which the trees are grown closely around the field. It may also serve as windbreaks. Mathur (1986) tried with *Prosopis juliflora* in Rajasthan for live-fencing. He opined that such fencing proved very effective and which was quite cheap also. On the same analogy, this alternative must be effectively tried in Ladakh region not only by the farmers but the Forest Department also. This will save a lot of money which can otherwise be used for some other developmental programmes.

11.3.2. Raising Trees with Cooperation

Cooperation from the bottom up is the essence of any programme. Same is true for afforestation works. If the farmers come together for such activities, then the programme becomes a grant success. In actual sense, such programmes are the programmes of the people. Therefore, every effort should be made to raise the plants without fencing. This will cut down the labour/time requirements. In Ladakh people are nature loving and only the need is to motivate them in the right direction for a noble cause. The tree species like *Populus* spp. and *Salix* spp. are very promising and fast growing and attain a good height in a short time and become out of reach of the browsing animals. Hence, such plantations can be raised without fencing without any danger of damages.

11.4. TACKLING WASTELANDS

The cold desert of Ladakh is very vast just like that of Rajasthan. The mountains are naked and there is always a danger of soil erosion. It has a big tract of wastelands which could otherwise be judiciously utilized by growing some suitable tree species or the grass which can provide a green cover. It is urgently needed to afforest the wastelands. This will not only give economy but also maintain the ecological balance.

11.5. RAISING LOCAL SPECIES

Many people advocate the idea of raising local species as they have some advantages over the exotics. But another school of thought opine that exotic species must be tried and if, they adopt under local conditions, they will prove beneficial. Although God has been unkind to this region (Ladakh) for giving scanty rains, it has been pleased to desert by giving some promising tree species which are cold resistant e.g. Poplars and Willows. Some exotic species like *Robinia pseudoacacia* can be introduced in the pastures and the plantations. Efforts should be made to motivate the farmers to grow these species on the agricultural farms.

The afforestation programme needs a continuous commitment from the Government and other agencies in view of the fact that forestry is a long spaned profession. However, some efforts are already in the direction. There is a great hinderance in the way that is the financial constraint. Therefore, bottom up plannings are always lacking. The power of people in bringing the revolution is well known the world over. If, the cold desert is to be saved from getting eroded, people should participate actively in the afforestation programme. The region is rich in man-power and the people have great love for the development of desert. If, the people contribute seriously, there is no reason why the constraint like financial should create hinderances in fulfilling the goals of the programme.

Therefore, there is an urgent need to formulate the policies for people's participation. However, these policies should be in accordance with physical, cultural, political and socio-economic environment of the region. Khan (1987) suggested the following methods for a successful participation in afforestation programme:

(a) tree tenure system,

(b) leasing of land to the landless rural families,

(c) organising tree farming cooperatives and leasing public lands to them,

(d) establishment of decentralised nurseries in the people's sector,

(e) organising village communities for afforestation and management of plantations on commons, and

(f) involvement of non-government organisation.

11.6. CONSTRAINTS

Afforestation work is not an easy task. Moreover, this if not impossible in the hilly tracts of cold desert but certainly is a difficult task to be accomplished. There are chances of failure of the programme due to one or the other reason. This could be attributed to some constraints in the way of afforestation works. Therefore, a critical analysis of the constraints is as follows:

11.6.1. Social and Economic

For a successful programme, the cooperation and involvement of the people (society) is very essential. It is expected that the maximum number of people should get the benefit of the project in the area. The land or capital is to be judged whether the same is useful or not i.e. what is the degree of its utility. It is a fact that people of Ladakh are tree conscious, what matter, it may be for their tangible benefits. Hence, the people socially and economically are not so strong. There is an urgent need to educate the people regarding their intangible benefits also.

11.6.2. Institution

The organisational and managerial capability is generally overlooked. This is a great constraint in the way of afforestation works. The success of any programme is attributed to the technology available with the local people. As is evident that afforestation is not an easy job and needs a high expertise. It requires a highly trained cruise. This will have better effects on the implementation of the project. Therefore, a net work of the developmental facilities must be created so that all types of projects can be run in the region for its overall development. There is an urgent need to establish a full-fledged research institute on the anology of Central Arid Zone Research Institute, Jodhpur (Rajasthan). This will cater the needs of the cold desert.

11.6.3. Technical

The technical constraints are the sophisticated restrictions on the methods employed in the region for its development. Following are the few technical constraints which are of much concern:

11.6.3.1. ***Landuse classification***: This indicates the maximum use of land. In the event of lacking of such classification, there is always confrontation between several projects. Hence, a detailed survey is the need of the day so that on the basis of information, future plans can be made well in time.

11.6.3.2. ***Restricted choice of species***: The adverse site conditions and expertise available restricts the choice of species for the region. The local vegetation, though a good guide for choosing the species but the number is very less. Moreover, only few species are able to resist the cold/winter which leads to the limited choice of species. Further, there is always a lacking of good sites of nursery purpose.

11.6.3.3. ***Use of available water***: The water sources of the region are very meagre. The only possibility is to store the precipitation, whatever, is available. But the problem is still aggravated because of the loss of water through evaporation. Hence, this loss is to be reduced as far as possible. The drip irrigation must be popularised in the area so that maximum use of water can be done, otherwise it is always a danger for the loss of human being (starving can be foreseen).

11.6.4. Labour/Financial

Labour availability in a project area is a vital factor. It has been noticed that labour is less as compared to the force of projects in cold desert. The labour availability becomes a severe constraint due to the urgency of agriculture. To tackle this problem, some economic devices must be developed for carrying out afforestation works without harming agriculture. For efficient work of afforestation, the ceiling on financial resources needs to be relaxed, or if, possible must be removed. For making the programme effective and efficient, comentoring and evaluation must also be carried out from time to time. This will ensure the right and judicious use of funds meant for developmental works/activities.

11.6.5. Improving Natural Habitat

Since fauna is an integral part of ecosystem in cold desert, so it can not be separated. Ladakh is blessed with a great number of fauna but very little has so far, been done for its improvement. It is now felt that to maintain a balanced ecosystem, protection and improvement of fauna needs immediate attention. Therefore, any planning for afforestation without inclusion of improvement of fauna is incomplete.

In conclusion, it is apparent that afforestation of cold desert is a challenging task. It can prove a bonanza for the improvement of ecosystem. Not only this, but the economic status of people can be alleviated. But it needs first priority in our national programmes. The results of afforestation can be visualised easily as it will ease the burden of forest which is of high economic value.

"Tree means Water,
Water means bread,
And bread is life".

—*K.M. Munshi*

Chapter 12

Agroforestry : New Approach

The multipurpose threes (MPTs) and fast growing nitrogen fixing trees (FGNFTs) have great potential to meet the needs of human-beings. They admittedly do good in the new subject areas of rural developmental forestry when a particular species is deliberately grown at one site and time to produce more than one product or benefit, but it is necessary to see them as only part of one or more land use systems designed to meet one or more objectives. The terminology for systems and objectives have become confused to the extent that agroforestry (defined in a group of systems) is generally equated with social forestry (defined in a group of objectives). These facts have, however, been discussed in more detail (Burley and Wood 1983).

In real sense, agroforestry is only one set of land management systems parallel to pure agriculture or pure silviculture systems. It has regenerated rather high levels of enthusiasm in recent years among researchers, developmental experts, and policy planners concerned with landuse systems. Although, it seems true that the scientific principles of agroforestry are only now being examined and, therefore, understood the practice in some form or other has been in existence since very early times (Khoshoo 1986) particularly among farmers in the Warmer parts of the World (Nair 1984). Accordingly, the efforts are now being made to devise the most appropriate ways to integrate the production of tree and other woody species with the production of agricultural crops and/or livestock simultaneously, from the same piece of land in a sustainable manner (Lundgren and Raintree 1982; Nair 1983; Torres 1983, 1984; Hoekstra 1985; Huxley 1985, 1986).

Agroforestry has most apparent potential in "marginal lands" and in resource - limiting small holding systems where monoculture (agriculture or forestry) may not be most feasible or desirable. This implies that agroforestry approach is based on the principles of "self-maintenance". Thus the systems and the management practices efficiency of inputs, whilst maintaining the productivity of the soil that supports, such a system is a matter of utmost concern in the present context. It is prudent to analyse the importance of agroforestry as a sound landuse system/approach in Ladakh region.

12.1 CREATION OF CONCEPT

This is, emphatically, not the practice of farming trees, a misconception that appears occasionally in North-American literature. However, elsewhere, in forestry literature, some of the observations and ideas, which contributed in creating agroforestry concept. Following are the few observations:

(a) Destruction of forests for the production of food and fuel and not for timber.

(b) Importance of perennial root systems to prevent nutrient loss through leaching and helped to design to emulate natural forest models.

(c) Room to establish soil fertility by slash and burn agriculture which led to design a forest/agriculture rotation.

(d) Live-fences are the age-old use of trees while farmers were pursuaded to put aside a portion of their land for fuelwood lots.

(e) Tree fodder feeds livestock in many areas of tropics and converted the trees in favour of shallow rooting ground fodder, and,

(f) Role of leguminous trees in traditional farming system. The leaves fall on to the ground as nitrogen - rich nutrients for the crops.

12.2 PRINCIPLES OF SELF-MAINTENANCE

The farmers have a number of demands which are directly related to forest. There are the needs upon which the farmers depend to achieve the principles of self-maintenance. These principles are the pillars for evaluating the economic status of the farmers. Two types of needs are recognised which are supposed to be meted out with agroforestry.

12.2.1. Primary Needs

These are the immediate needs conceptualized in the concept of 5Fs as follows:

(i) Food,

(ii) Fuel,

(iii) Fodder,

(iv) Fertilizer, and

(v) Fibre.

12.2.2. Secondary Needs

These needs can be meted out with over the pace of time. A few of the needs are as follows:

(i) *Nitrogen Fixation*: The most common advantage attributed tc agroforestry is the possibility for the improvement of fertility status of agricultural lands through additional amounts of nitrogen attributed by the tree legume component. Although, the legumes are not the only group of woody species that have a potential role of agroforestry, the family leguminosae offers by far the greatest choice of woody species for agroforestry in terms of economic uses as well as ecological adaptability. Therefore, among the various avenues of nitrogen addition to soil through natural or biological means, the most significant one that is brought about by the presence of trees on agricultural land could be nitrogen fixation by leguminous trees.

(ii) *Soil Conservation*: The potential role of trees in reducing run-off and erosion losses is well apprehended and understood but not well documented. The physiognomy of natural forest communities is such that it provides a multilayer defence against the impact of hostile environmental conditions. The litter and humic layers on the soil surface act as a cushion. The soil conservation benefit of woody perennials can conveniently be made use of agroforestry in several situations.

(iii) *Mulch Farming* : Mulching with crop residues and other vegetative matter has been a consistently successful practice. Several advantages have been attributed to the practice of adding mulch (Lal 1973 and 1975). Moreover, presence of organic matter on soil also improves physical conditions (Lal 1979) and fertility status of soil (Abuzaid 1973). These principles and practices of mulching are very relevant to agroforestry particularly in degraded and waste lands. The hedge-row intercropping (alley cropping) system is an excellent possibility for this region.

(iv) *Windbreaks and Shelterbelts*: Ladakh region experiences moderate to very strong winds, which blow continuously, can have direct and indirect harmful effects on the plants. The use of trees and other woody perennials to protect agricultural fields is a widely adapted practice in agricultural systems. The principle can be a considerable use in evolving sound agroforestry technology for such areas. However, selection of agroforestry species will depend upon the expertise and choice of woody perennial to be used as windbreaks. Regarding this, very promising results have been reported from the studies conducted at PFRI, Peshawar (Sheikh and Chima 1976; Sheikh and Khalique 1982).

12.3. EVOLVING APPROPRIATE TECHNOLOGY

A preliminary evaluation of potential agroforestry and its systems as well as non-agroforestry technologies recognised for the region, which can be a starting point for more indepth analysis in future, is given as under:

12.3.1. Non-Agroforestry Interventions

The major areas of agroforestry interventions are identified as soil and water conservation, maintaining and improving the available soil nutrients, improving fodder quality and quantity etc. Agroforestry technologies proposed are not the solutions and hence alternative non-agroforestry interventions need to be considered. Following could be the general observations on such interventions:

12.3.1.1. ***Use of Cattle dung as manure*** : For increasing the available soil nutrients, the animal manure may be used. It has been generally recorded that 5-10 t ha^{-1} yr^{1} is very beneficial to many crops, but in most of the farming systems, the supply is not likely to exceed 1-2t ha^{-1} with a production of about 1.0 t adult $animal^{-1}yr^{1}$. Therefore, this practice holds good in Ladakh region. However, it is/will be effective on small scale.

12.3.1.2. ***Use of artificial fertilizers***: Though, it is a well established fact that artificial fertilizers increase the yield of several crops, yet the farmers seem to heritate to adopt this practice frequently. This may be due to infancy of artificial fertilizer programmes or more structural problems (e.g. price hike), it yet to be analysed. However, it may be realistic to assume that the use of artificial fertilizer can play a vital role in alleviating some of the fertility problems of the area.

12.3.1.3. ***On farm bunding***: It is an observation that the State Government will have to embark on a programme of soil and water conservation in marginal lands. A comprehensive cost/benefit ratio is to be taken into consideration on priority and to see whether cold arid land development is financially feasible on large scale.

12.3.2. Agroforestry Interventions

Besides non-agroforestry interventions, there are some agroforestry interventions also. The following are some of them:

12.3.2.1. ***Trees/shrubs for timber and fodder***: The farmers of cold desert have already taken up this practice on a limited scale and at minimal input by making use of saplings of Willows (*Salix* spp.) and Poplars (*Populus* spp.) on the bunds. However, the bottlenecks to increase the number of trees on bunds seem to be the protection issue and interference with crop production because of competition for land and labour. Moreover, analysis must be focus on the trade-off between crop production and tree production.

12.3.2.2. ***Block rotational planting***: This system is yet unknown and gets fit in marginal lands in the first instance, if it costs in terms of annual crop production but would not be as high as that of productive land. Hence, the hardworkers but the poor resource the farmers will not be able to cope up with reduction in their farm income. However, temporary renting by the Government may bring the system within the reach of resource-poor land tillers.

12.3.2.3. ***Agri-silvicultural system***: This system would be production oriented so that the tree product compliment the present annual crop component. The farmers would only be prepared to make such charges, if, the return to the farm increase and/or are more secure and/or better distributed in comparison to present annual cropping (monoculture) systems. This is, however, likely to be achieved for the resource-poor farmers by maximising the returns per unit area at the expense of additional labour, while the returns to the resource-rich farmers are likely to increase by maximising the returns of labour.

12.3.2.4. ***Popularising sylvo-pastoral system***: This system is fairly a radical change from the present annual cropping system and it is likely to reduce the labour input on the per unit area basis, possibly with an increase in the returns per labour. More than 50 per cent of the land area comes under cultivable wastes and barrens which have maximum potential for this system. Poplars and Willows (silvicultural elements) would be used alongwith leguminous grass elements which are found in abundance in the region.

12.3.2.5. ***Micro-zonal planting system***: The services like soil and water conservation, production (fodder, mulch and fuel) under agroforestry have been derived from its tree components. Evaluation of system should focus on trade-off between crop production and tree products (mainly the fodder) in terms of income (including security and distribution), labour utilization and local food conservation benefits of this system, not only its initial benefits but also the development over time should be studied.

12.3.2.6. ***Multipurpose trees in nullahs and commons***: The issue at stake with this agroforestry intervention is the access to products by the landless as well as employment opportunity resulting from this activity (plant propagation, site preparation, planting, weeding, harvesting of the products for homeuse, sale or processing). Although, plant resources in woody species of multipurpose utility in this system are rather meagre. Already introduced trees of poplars, willows, mulberry, and walnuts, can easily be exploited. The patential management vis-a-vis economic feasibility of the benefits will have to be worked out.

12.4. PERSPECTIVES OF AGROFORESTRY

There is a great scope for research and development programmes in the region. For its development following actions may be taken up immediately.

12.4.1. Evolving System

It has been debated a number of times that agroforestry can cater the needs of the people. Since, Ladakh consists of the largest area, so there is an ample scope of agroforestry. All the systems (traditional and new ones) should be tried and the most beneficial ones should be evolved. Sagwal and Amani (1986) have suggested the research needs and extension approaches/activities for marginal lands. Therefore, these suggestions may also be followed for strengthening the approaches of agroforestry (the relevant systems).

12.4.2. Rearing Wildlife

Agroforestry is a system which improves the environment and keeps a blanace in the ecosystem. There is an urgent need to undertake some researches on wildlife as it forms an integral part of the forest (artificial or natural). Therefore, artificial forests must be raised depending upon the suitability and adaptability of the tree species and wildlife can be reared easily under such conditions.

12.4.3. Promoting Forestry Extension

It is the way to educate the people about forestry and its components. Through this programme, the planning of afforestation works should be framed. Therefore, whole land potential suited to raise the plantation must be assessed by the help of detailed survey of the region. Negi (1985) outlined the areas for such programmes in dry temperate Himalaya which can be the base for the agroforestry programme.

12.4.4. Exploration of MFP

The region has a great potential of MFP (Minor Forest Products), mostly the medicinal and aromatic plants. Sagwal and Bali (1987) have suggested the possibility of exploration of this wealth in Ladakh region. Based on the same lines, the programme can be chalked out for the development of Minor Forest Products which can boost the income of the farmers and agroforestry can play a crucial role in this direction.

12.4.5. Establishing Research Institute

The cold desert of J & K State is of a unique type and needs immediate attention for its development. A full-fledged Research Institute for forestry particularly the agroforestry should be established to cater the research needs of the areas. This can match the model of Central Arid Zone Research Institute, Jodhpur (Rajasthan), and Centre for Agroforestry at Jhansi (U.P.). An initiative should be taken up from the Top Down Planning so that it may give a right direction in the overall development.

12.4.6. Other Suggestions

In addition to above suggestion, there may be the following points to be taken up with the pace of time:

(i) setting up of forest forums/clubs,

(ii) framing a realistic landuse policy,

(iii) provision of bottom up planning,

(iv) establishing small scale industries etc.,

(v) women's participation in development activities.

12.5. CONSTRAINTS OF AGROFORESTRY

A cursory glance over the landuse and cropping pattern in cold desert (Ladakh) had revealed that in view of the extreme hostile climatic conditions, low soil productivity and very short growth period with predominant mono-cropping system, neither agriculture nor forest along can meet out the minimal needs of the people. Primarily agriculture and secondarily lands put to other uses (multiuse land resources) will have to be the main stay and this alone can save the situation from bad to worst. However, in ultimate analysis mixed and diversified production on agricultural land (i.e. agroforestry) is the only answer, besides extensive plantations fodder on waste/barren and the marginal lands.

However, to set the fact straight, it is conceded that agroforestry on agricultural lands and land not suited to intensive agriculture herein referred to as marginal and waste/barren lands has not received much attention of researchers primarily due to our pre-occupation to increase food production, which has been in short supply in the recent past. It is also possible that there has been a general feeling among the farmers and other agriculture/forest managers that trees compete with agricultural crops for water, plant nutrients and sun-shine and also give shelter to crop pests and the birds.

In the light of the above remarks, it would be germane to identify pitfalls/constraints hindering universal acceptance of agroforestry practices by the farming community and also retarding the pace of reseach activities. Some of the major constraints of agro-forestry pratices and research recognized are as follows:

12.5.1. Biological

Due to prolong and severe winter regimes in Ladakh region, plants/trees are left with 'limited growth period'. Further, it has been noticed that farming community just plant trees and try to *domesticate* them under adverse growth conditions which lead to several weak links of scientific thinking. Therefore, lack of *intensive management* in the initial stages of plantations fails the farmers to reap the rich products (timber, fuel, fodder etc.) as intended. Besides these, failure of the farmers to take stock off *natural hazards*, e.g. damping-off of seedlings of many tree species and frost bite, and also damages caused by insect-pests pose a severe set-back in proper growth and development to tree and agricultural crops.

12.5.2. Socio-economic

As earlier stated more than 78 per cent of land holdings fall between 0.5 and 2.0 ha. in Ladakh, giving a very pessimistic and deplorable condition, it will be like asking too much from the farmers to plant trees in their land for two basic reasons-food-being the first priority and any reduction in their expected produce will cause them heavily financially, further due to prolong gestation period of tree crops, the resource-poor farmers will not be able to cope with reduction in income during early development, i.e. reduction in income coupled with *no immediate gains*. Besides, several religious sentiments and superstitions attached to tree plantation and its ultimate harvest, pose a severe threat in adopting agroforestry practices.

12.5.3. Agro-eco-climatological

Because of great diversity in climatic regimes (rain, temperature etc.), soil and topography, it does not auger well while practising both forestry and agriculture. However, the weather being absolutely unpredicted frequent failure of crops is inevitable, besides causing great set back to artificial regulation of tree crops. The growth period being very short, mostly mono-cropping practices are in vague in the region and that too in a very small landholding, thus providing less opportunities for planting trees in agriculture crops except on the bunds. For most of the period of the year, the cultivable lands and other waste lands are exposed to the onslaught of hostle weather conditions causing soil and water erosion and hence reducing the fertility and productivity of soil and lowering its water regimes.

12.5.4. Management/Institutions

Lack of necessary infrastructure and scientific manpower by state Government is a major constraint because continuous research efforts and innovative mechanisms are of utmost importance to cater the growing needs of the farmers/foresters. Besides, inadequate extension base causes a perpetual constraints. Further, selection of good seed/planting material and its availability creates hinderances in the proper development of agroforestry in the region.

It is concluded that agroforestry has ample scope in the region. For achieving the principles of self-maintenance, sincere efforts must be done. All the possible attributes of non-agroforestry and agroforestry interventions must be exploited. There are still more chances for the development of the area and hence the suggestions for R and D programme should be earnestly tried for their feasibility.

Bibliography

Abu-Zaid, M.O. 1973. Continuous cropping in the areas of shifting cultivation in Southern Sudan. Trop. Agri. (Trinidad), 50:285-90.

Anonymous. 1972. Indian Forest Utilization. Vol.II. The Manager of Publications. Delhi. 941 pp.

Anonymous. 1977. Plan of action to stop desertification. UNCOD A/ CONF 74/36, Report on UN Conf. on Desertification, Nairobi (Kenya).

Anonymous. 1983. National Agricultural Research Projects (ICAR/ IBRD Projects). University Background Paper. S.K. University of Agricultural Sciences & Technology, Jammu and Kashmir State, October, 1983.

Anonymous. 1985. The Vun. Jammu and Kashmir Forest Department, Srinagar (J&K State), 8(1):1-70.

Anonymous. 1988. Greening the Desert. Desertification Control Bulletin. 17:50.

Bailey, J.A. 1984. Principles of Wildlife Management. John Wiley & Sons. New York. 373 pp.

Bhat, G.M. 1965. The Soils of Kashmir. Bull. Agri. Department. J&K Government, Jammu and Kashmir State, India.

Buchanan, R.O. 1959. Some Reflections on Agricultural Geography. Geography. 44: 1-2.

Burley, J. and Wood, P.J. 1983. Development of Curricula for Community forestry. Pap. Presented 12th Session, FAO, Adv. Comm. Forestry Education, Nairobi, Kenya.

Burrand, S.G. and Heydon, H.H. 1907. A sketch of geography and geology of Himalayan Mountains and Tibet, Calcutta, India.

California Fertilizer Association. 1985. Western Fertilizer Handbook. Seventh Edition. The Interstate Printers & Publishers, Inc., Illinois (U.S.A.). 288 pp.

Caughley, G. 1977. Analysis of Vertebrate Populations. John Wiley and Sons. New York. 234 pp.

Chadha, S.K. 1990 a. The impact of Agriculture on Environment in Kashmir. In: S.K. Chadha (ed.). Ecology of Kashmir. Ashish Publishing House. New Delhi-26, 211 pp.

Chadha, S.K. 1990 b. The Environmental Challenges in Kashmir Himalaya. pp. 131-36. In: S.K.Chadha (ed.). Ecology of Kashmir. Ashish Publishing House. New Delhi-26, 211 pp.

Champion, H.G. and Seth S.K. 1968. General Silviculture for India. Government of India Publication Branch. Delhi. 511 pp.

Chopra, P.N. 1980. Ladakh. S.Chand and Company Ltd. Ram Nagar, New Delhi-55, 109 pp.

Dar, K.M. 1986. Constraints in Agriculture Production. Proc: Seminar on Constraints in the Agriculture Production in Jammu and Kashmir State, S.K. University of Agricultural Sciences & Technology, Shalimar Campus, Srinagar (J & K State), India. July 8-9, 1986.

Dregne, H.E. 1976. Desertification: Symptom of a crisis. Quoted in Handbook of Desertification Indicators based on the Science Association's Nairobi Seminar on Desertification (ed.: P. Reining). AAAS, Washington, D.C., 141 pp.

Durani, P.K., Singh, G. and Kachroo, P. 1975. Phytosociological studies on the vegetation of Ladakh Desert. Ann. Arid Zone. 14(2): 75-86.

Ehrlich, P.R. and Ehrlich, A.H. 1970. Population, Resources, Environment. W.H. Freeman and Co., San Franchs Co., U.S.A. 383 pp.

Goodall, J. 1965. Chimpanzees of the Gombe Stream Reserve. pp. 425-73. In: Primate Behaviour, I. Devore (ed.). Holt, Rinchart and Winston, Inc. 654 pp.

Gupta, M.P. 1983. Role of forestry extension education in Himachal Pradesh. Proc: Nat. Symp. Improvement of Forest Biomass. pp. 323-27. In: P.K. Khosla (ed.), ISTS. Solan (H.P.).

Gupta, K.R. 1986. Role of Soil and Water Conservation in the Development of Cold Arid Desert of Ladakh. Proc: National Seminar-cum-Workshop on Development of the Cold Desert of Ladakh. The Leh Desert Development Agencies, September 23-25, 1986.

Handoo, G.M. and Associates. 1984. Pre-irrigation Soil Survey of Proposed Command Area of Igophey Canal Project Leh (Ladakh). Soil Survey Organisation Department of Agriculture, J&K State, India.

Hassnain, F.M., Masato, C.K.I., and Tokan, S.D. 1977. Ladakh - The Moonland. Light and Life Publishers. New Delhi, India.

Hellden, U. 1988. Desertification Monitoring : Is the Desert Encroaching? Desertification Control Bulletin. 17:8.

Hoekstra, D.A. 1985. Economic Concepts of Agroforestry. ICRAF, Nairobi. p. 12.

Huxley, P.A. 1985. Systematic Designs for field experimentation with multipurpose trees. Agroforestry Systems. 3:197-207.

Huxley, P.A. 1986. Tree/Crop Interface or simplifying the Biological Environmental Study of mixed cropping. Agroforestry Systems, 3:251-56.

Kangoo, G.H. 1989. Soil Conservation in Jammu and Kashmir: Problems and Prospects. Forest News. Planning and Publicity Division. Jammu and Kashmir Forest Department. Srinagar (J&K State), 2(4):1-9.

Khan, I. 1987. Westelands Afforestation: Techniques and systems. Oxford & IBH Publishing Co. Pvt. Ltd. New Delhi - 1, 176 pp.

Khan, G.M. and Wani, M.Y. 1986. Soil and Water Management in increasing productivity of crops in cold Arid Region of Ladakh. Proc: National Seminar-cum-Workshop on Development of Cold Desert of Ladakh. Leh Desert Development Agencies, September 23-25, 1986.

Khoshoo, T.N. 1986. Environmental Priorities in India and sustainable Development. 73rd Session. Indian Science Congress Association, New Delhi, pp. 61-62.

King, R.T. 1966. Wildlife and Man. New York Conservationist. 20:8-11.

King, K.F.S. 1978. Agroforestry. Paper presented to the 50th Agricultural Conference. Amsterdam.

Lal, R. 1973. Effects of seedbed preparation and time of planting on maize in Western Nigeria. Expl. Agri., 9:303-13.

Lal, R. 1975. Role of mulching techniques in Tropical Soil and water Management. IITA. Tech. Bull. No. 1, IITA. Ibadan, Nigeria, p. 38.

Lal, R. 1979. Effects of Cultural and harvesting practices on Soil physical conditions. In: Mongi, H.O. and P.A. Huxley (eds.), Soils Research in Agroforestry, Nairobi, ICRAF. pp. 327-61.

Leopold, A. 1949. A Sand County Almanac. Oxford University Press. New York. 226 pp.

Lundgren, B.O. and Raintree, J.B. 1982. Sustained agroforestry in Agricultural/Research for Development: Potentials and Challenges in Asia. Report Conference (24-29th October, 1982), Jakarta, Indonesia, The Hague, ISNAR, pp. 37-49.

Mann, H.S. 1982. Desertification in India: Past and Future of the Thar Desert. pp. 44-45. 12th International Congress of Soil Science. New Delhi, India.

Mathur, C.M. 1986. Plantation for fuel, fodder and small timber in Rajasthan Desert. Proc: Fourth All India Conference on Desert Technology. Indian Society of Desert Technology, Jodhpur (Rajasthan), India.

Mathur, N.K. and Sharma, A.K. 1983. West Himalayan Ecology and Agroforestry. Paper presented at National Workshop on Agroforestry. Government of India and Haryana State. National Dairy Research Institute Karnal (Haryana). July 23-25, 1983.

Misri, B. 1980. A Preliminary Survey of grasses and legumes - Review of Ladakh. Australian Plant Introduction, USIRO, 30:37-45.

Mukhopadhyay, A. 1971. Agricultural Extension - A Field Study. The Minerva Associates. Calcutta. 233 pp.

Murthy, R.S. and Pandey, S. 1978. Soil and Landuse in Himalayan Region. Nat Beau. of Soil Survey and Landuse Planning, New Delhi. pp. 9.

Nair, P.K.R. 1983. Tree Integration in farm lands for sustained productivity on small holdings. Environmentally Sound Agriculture. Praeger Publishers, New York. pp. 315-33.

Nair, P.K.R. 1984. Soil Productivity Aspects of Agroforestry. Science and Practice of Agroforestry. ICRAF, Nairobi, Kenya. p. 49.

NCA. 1976. National Commission on Agriculture Report of Ministry of Agriculture. Part IX-Forestry. Government of India, New Delhi, 457 pp.

Negi, S.S. 1983. Forest Protection. Foundamentals of Forestry. Vol. X. Bishen Singh Mahendra Pal Singh. Dehradun - 1(India). 103 pp.

Negi, J.P. 1985. Afforestation of Dry Temperate Himalayas. Lecture (special) delivered at National Symposium on "Production and Conservation Forestry". ISTS, H.P.A.U., Solan, April, 1985.

Pathak, S. 1958. Farm Forestry in India. Proc: Farm Forestry Symposium, Dehradun. ICAR, New Delhi-1.

Press Trust of India. 1991. Giving Tree Cover to Ladakh region. The Tribune, Saturday, April 27, 1991.

Rapp, A. 1974. Review of Desertization in Africa—Water, Vegetation, and man. Reprint of SIES Report No. 1, 1974. In: Lunds Universitets Naturgeografiska Institution, Reporter Och Notiser, 39:77.

Reddy, M.P. 1983. Dwindling Forests pose Threat. Trees pay well in Marginal Lands. 83 pp.

Sagwal, S.S. 1981. All about Forestry Extension. Indian Farmer's Digest. 10(1) : 17-18.

Sagwal, S.S. 1985. Amenity Plantations: for health, Wealth and beauty. Farmers' Journal. September, 1985. 5(5): 48-51 & 53.

Sagwal, S.S. 1991. Environmental Pollution in Himalayas - A case History of Jammu and Kashmir State. pp. 51-61. In: S.K.Chadha (ed.). Soiled Splendour of Himalaya, Vinod Publishers & Distributors, Jammu (India). 151 pp.

Sagwal, S.S. 1991. The Wildlife in Ladakh - An Ecological porspective, pp. 85-91. In: S.K.Chadha (ed.). Soiled Splandour of Himalaya. Vinod Publishers & Distributors. Jammu (J & K), India. 151 pp.

Sagwal, S.S. and Amani, A.Z. 1986. Afforestation of Marginal Lands with special reference to Jammu and Kashmir State. pp. 99-104. In: H.C.Srivastava, B. Vatsaya, and K.K.G.Menon. (eds.). Proc: Symposium on "Plantation Opportunities in India". Vol. I. Oxford & IBH Pub. Co., New Delhi - 1, 411 pp.

Sagwal, S.S. and Bali, A.S. 1987. Exploration of Medicinal and Aromatic Plant Wealth of Ladakh region of Jammu and Kashmir State. Ind. Perfumer, 31(4): 340-46.

Sagwal, S.S. and Gupta, N.K. 1988. Role of Forestry Extension in Exploitation of Medicinal and Aromatic Plants of Himachal Pradesh. In: P.K. Khosla and R.N. Sehgal (eds.) Trends in Tree Sciences. ISTS. pp. 50-54.

Sahni, Kamal. 1990. The Animal Husbandry in Ladakh - An Ecological Perspective. In: S.K. Chadha (ed.). Ecology of Kashmir, Ashish Publishing House New Delhi - 26, 211 pp.

Salaria, S.A. 1979. Forest Wealth in Jammu and Kashmir. In: Krishna Murti Gupta and Desh Bandhu (ed.). Man and Forest. Today & Tomorrow's Printers & Publishers, Karol Bagh, New Delhi-5, 329 pp.

Schaller, G.B. 1972. Predators of Serengeti: Part 2, Are You running with me, Huminid Natural History. 81: 60-69.

Schemnitz, S.D. 1980. Wildlife Management Techniques Manual (ed.). 4th ed. The Wildl. Soc., Washington, D.C. 686 pp.

Sethi, C.M. 1977. Mobilization of youth for Social Forestry. Indian Farming. 26(2): 103-4.

Shah, M.H. 1986. Constraints in Fodder Production in J&K State. Proc: Seminar on Constraints in the Agriculture Production in Jammu and Kashmir State, S.K. University of Agricultural Sciences & Technology, Shalimar Campus, Srinagar (J&K State), India. July 8-9, 1986.

Sheikh, M.I. and Chima, A.M. 1976. Effects of Windbreaks (trees) rows on the yield of wheat crop. Pak. J. Forestry. 32:21-23.

Sheikh, M.I. and Khalique, A. 1982. Effects of tree belts on the yield of Agricultural Crops. Pak. J.Forestry. 38: 21-23.

Shrivastava, A.K. 1989. People's Participation in Forestry. Forest News. Planning and Publicity Division, Jammu and Kashmir Forest Department, Srinagar (J&K State). 2(5):1-6.

Singh, S. 1970. Minor Forest Produce in Ladakh - A Big Potential. Proc: State Forestry Conference, J & K State, October 12-16, 1970. J & K State Forest Department, Srinagar. 311 pp.

Singh, R.V. 1977. Management of Natural grasslands in the Himalayas for Protein Production. Himalaya, Man, and Nature. Vol. 1, No. 7, 1977, Dehradun (India).

Singh, K.P. 1981, Methodology T & V System: Implimentation in Haryana State. Krishi Agradoot. Jan. June, 1981, pp. 16-19.

Steinhoff, H.W. 1971. Communicating complete wildlife values of Kenai. Trans. N. Amer. Wildl. and Nat. Res.. Conf. 36:428-38.

Talib, A.R. 1986. Soil and Water Conservation Measures to Combat Desertification in Cold Arid Region of the Himalayas. Proc: National Seminar-cum-Workshop on development of Cold Desert of Ladakh, Leh. Desert Development Agencies. September 23-25, 1986.

Tejwani, K.G. 1974. Case Studies. On Farming Systems in the Semi and Tropics of India. pp. 141-56. International Workshop on Farming Systems. 1982. ICRISAT, Hyderabad (A.P.), India.

Jejwani, K.G. and Bhardwaj, S.P. 1982. Soil and Water Conservation Research. pp. 608-21. Review of Soil Research in India (Part-II). 12th International Congress of Soil Science. New Delhi.

Torres, F. 1983. Role of Woody Perennials in Animal Agroforestry. Agroforestry Systems. 1:131-63.

Torres, F. 1984. Potential Contribution of Leucaena hedge-rows Intercropped with Maize to the production of organic Nitrogen and Fuelwood in the low-land Tropics. Agroforestry Systems. 1:323-25.

Trigotra, R.C. 1983. Social Forestry and need for motivation and

Extension. Paper presented at the National Workshop on Agroforestry. NDRI, Karnal (Haryana). July 21-23, 1983.

UNCOD A/CONF 74/2. 1977. World Map of Desertification. DOC. of the UN Conf. on Desertification, Nairobi (Kenya).

Vijaylakshmi, K. 1986. Land classification, Landuse Systems, and Agroforestry. Paper presented at Workshop/Training Course on Agroforestry Research. Hyderabad (A.P.). India. ICAR & ICRAF, 16th September-1st October, 1986.

Wani, M.Y. and Associates. 1977. Soils of Cold Arid Zone. Soil Testing Laboratory. Leh (Ladakh), J&K State, India.

Index